Agile Machine Learning with DataRobot

Automate each step of the machine learning life cycle, from understanding problems to delivering value

Bipin Chadha

Sylvester Juwe

BIRMINGHAM—MUMBAI

Agile Machine Learning with DataRobot

Publishing Product Manager: Sunith Shetty
Senior Editor: Mohammed Yusuf Imaratwale
Content Development Editor: Nazia Shaikh
Technical Editor: Devanshi Ayare
Copy Editor: Safis Editing
Project Coordinator: Aparna Ravikumar Nair
Proofreader: Safis Editing
Indexer: Hemangini Bari
Production Designer: Sinhayna Bais

First published: December 2021

Production reference: 1191121

Published by Packt Publishing Ltd.
Livery Place
35 Livery Street
Birmingham
B3 2PB, UK.

ISBN 978-1-80107-680-7

www.packt.com

This book is dedicated to my father Satdevraj Chadha and my wife Madhumita Chadha, who are the inspirations for this work – Bipin Chadha

This book is dedicated to my family and close friends who constantly supported and encouraged me during this project - Sylvester Juwe

Contributors

About the authors

Bipin Chadha is a hands-on leader of data science teams who can find innovative solutions to complex problems. He creates systemic data-driven models that enable executives to understand how their business operates, analyze a broad range of scenarios and strategies, and understand the likely implications of decisions and events prior to implementing risky changes. His passion is to build data-driven cultures, develop effective teams, and drive organizations to grow and succeed.

Sylvester Juwe is a highly accomplished executive, with hands-on technical expertise in implementing complex big data and advanced analytics solutions, from conceptualization to the commercial impact. He is a well-versed leader who leverages sophisticated data capabilities, influences stakeholders, and creates a strong culture of governance and curiosity in solving complex business challenges, thereby creating a commercial impact.

About the reviewer

Aman Sharma is a senior data scientist at DataRobot. Aman has a background in computer science and has worked in various industries as a full stack data scientist. He has extensive experience in demand forecasting, propensity, churn, credit risk, fraud, marketing attribution, and optimization use cases. He works at DataRobot, which is a leading end-to-end enterprise AI platform.

Table of Contents

3

Understanding and Defining Business Problems

Section 2: Full ML Life Cycle with DataRobot: Concept to Value

4

Preparing Data for DataRobot

Section 3: Advanced Topics

9

Forecasting and Time Series Modeling

10

Recommender Systems

11

Working with Geospatial Data, NLP, and Image Processing

Preface

DataRobot enables data science teams to become more efficient and productive. This book helps you address **machine learning** (**ML**) challenges with DataRobot's enterprise platform, enabling you to extract business value from data and rapidly generate commercial impact for your organization.

You'll begin by learning how to use DataRobot's features to perform data prep and cleansing tasks automatically. The book covers best practices for building and deploying ML models, along with challenges faced while scaling them to handle complex business problems. Moving on, you'll perform **exploratory data analysis** (**EDA**) tasks to prepare your data to build ML models and ways to interpret results. You'll also discover how to analyze the model's predictions and turn them into actionable insights for business users. After that, you'll create model documentation for internal as well as compliance purposes and learn how the model gets deployed as an API. In addition, you'll find out how to operationalize and monitor the model's performance. Finally, you'll work with examples of time series forecasting, NLP, image processing, MLOps, and more using advanced DataRobot capabilities.

By the end of this book, you'll have learned how to use some of the AutoML and MLOps features DataRobot offers to scale ML model building by avoiding repetitive tasks and common errors.

Who this book is for

This book is for data scientists, data analysts, and data enthusiasts looking for a practical guide to building and deploying robust ML models using DataRobot. Experienced data scientists will also find this book helpful for rapidly exploring and building and deploying a broader range of models. The book assumes a basic understanding of ML.

What this book covers

Chapter 1, *What Is DataRobot and Why You Need It*, describes the current practices and process of building and deploying ML models, and some of the challenges in scaling that approach. This chapter will then describe what DataRobot is and how DataRobot addresses many of these challenges, thus allowing analysts and data scientists to quickly add value to their organization. This also helps executives understand how they can use DataRobot to efficiently scale their data science practice without a need to hire a large staff with hard-to-find skills. This chapter also describes various components of DataRobot, how it is architected, how it integrates with other tools, and different options to set it up on-premises or in the cloud. It also describes, at a high level, various user interface components and what they signify.

Chapter 2, *Machine Learning Basics*, covers some basic concepts of ML that will be used and referenced in this book. This is the bare minimum you need to know to use DataRobot effectively. It is not the intent of this chapter to give you a comprehensive understanding of ML, but just a refresher of some key ideas.

Chapter 3, *Understanding and Defining Business Problems*, will show you examples of how to get to the root of a problem and then set it up as an ML project. A business problem needs to be carefully defined and turned into an ML problem for it to be solved with DataRobot. This is a critical step that is often ignored, resulting in problems and failures downstream. Please review this chapter carefully to prevent the wastage of a lot of hard work. This chapter is tool- and ML method-agnostic.

Chapter 4, *Preparing Data for DataRobot*, covers how to stitch data together from multiple disparate sources at a high level. Depending on the data, DataRobot might perform data prep and cleansing tasks automatically, or you might have to do some of these on your own. This chapter covers concepts and examples to show how to cleanse and prepare your data and the features that DataRobot provides to help with these tasks.

Chapter 5, *Exploratory Data Analysis with DataRobot*, will show you how to use DataRobot to perform various data analyses and get data ready to start building models. We provide detailed examples of the kinds of analysis that should be done and what to be aware of to prevent issues downstream. Done right, this analysis can help catch data problems and also generate useful business insights.

Chapter 6, *Model Building with DataRobot*, shows step-by-step examples of building different types of models with DataRobot. We cover details such as what settings to use under different circumstances, how to select specific model types, setting up cross validation, building ensemble models, and tracking the top-performing models on the leaderboard.

Chapter 7, Model Understanding and Explainability, will show you examples of various functions and outputs that DataRobot provides to help you understand the models and select the one that best solves the business problem. In this chapter, we will cover, via examples, what aspects you need to watch out for, and the trade-offs you have to make in model selection.

Chapter 8, Model Scoring and Deployment, covers how to use models to score input datasets, create predictions to be used in the intended applications, deploy models in production, and monitor models.

Chapter 9, Forecasting and Time Series Modeling, describes how you go about building time series models. These types of models are typically used for forecasting applications. The chapter shows examples of how different time series problems are handled with DataRobot. We cover single- as well as multi-series problems.

Chapter 10, Recommender Systems, covers examples of how you go about building recommender systems with DataRobot. These types of models are typically used for recommending products or services to users. The chapter covers the strategies and functionality differences in how a recommendation problem is handled with DataRobot. We cover trade-offs associated with building different recommender models.

Chapter 11, Working with Geospatial Data, NLP, and Image Processing, covers various DataRobot functions relating to visualization and analysis of geospatial, text, and image features, as well as building ML models that incorporate such features. This chapter describes DataRobot capabilities to automatically incorporate text and image data into ML models, thereby improving the performance of these models.

Chapter 12, DataRobot Python API, describes when and how to use the DataRobot Python API. While DataRobot automates many aspects of model building, there are many scenarios where you need to use programming languages such as Python to efficiently and scalably perform ML tasks. DataRobot provides a convenient API that allows experienced data scientists to execute DataRobot functions programmatically.

Chapter 13, Model Governance and MLOps, covers some recent topics that are beginning to get a lot of attention. Once a model has been developed and deployed, it needs to be governed and maintained over time. While this is similar to an IT system in many ways, there are some critical differences that need to be understood and operationalized. This chapter covers several features and functions that DataRobot provides to assist in governing and maintaining ML models.

Chapter 14, Conclusion, covers where to go for additional information and other topics that might be outside the scope of this book. We also describe where we see automated ML and DataRobot heading in the future.

To get the most out of this book

To get the most out of this book, you will need access to the DataRobot software. The commercial version has all the functionality. If you do not have access to a commercial version, you can get an evaluation version that works for a limited time and does not have all the capabilities discussed. For some of the advanced capabilities and the API to work, you will need to know some Python and have access to an open source Python environment (for instance, Anaconda or Jupyter Notebooks).

Software/hardware covered in the book	Operating system requirements
DataRobot	Windows, macOS, or Linux

Even though most of what we describe in this book can be done without knowing Python, we highly encourage you to learn Python as a next step. Knowing programming languages such as Python will open up a lot more possibilities for you and enable you to take better advantage of tools such as DataRobot.

Code in Action

The Code in Action videos for this book can be viewed at `https://bit.ly/3cj2qp1`.

Download the color images

We also provide a PDF file that has color images of the screenshots and diagrams used in this book. You can download it here: `https://static.packt-cdn.com/downloads/9781801076807_ColorImages.pdf`.

Conventions used

There are a number of text conventions used throughout this book.

`Code in text`: Indicates code words in text, database table names, folder names, filenames, file extensions, pathnames, dummy URLs, user input, and Twitter handles. Here is an example: "For our purposes, we simply created a copy of our `imports-85-data.xlsx` dataset file and named it `imports-85-data-score.xlsx`."

A block of code is set as follows:

```
deployment = dr.Deployment.create_from_learning_model(
    MODEL_ID, label='DEPLOYMENT_LABEL',
    description='DEPLOYMENT_DESCRIPTION',
    default_prediction_server_id=PREDICTION_SERVER_ID)
deployment
```

Bold: Indicates a new term, an important word, or words that you see onscreen. For instance, words in menus or dialog boxes appear in **bold**. Here is an example: "After selecting the options, we can click on the **Compute and download predictions** button."

> **Tips or Important notes**
> Appear like this.

Get in touch

Feedback from our readers is always welcome.

General feedback: If you have questions about any aspect of this book, email us at customercare@packtpub.com and mention the book title in the subject of your message.

Errata: Although we have taken every care to ensure the accuracy of our content, mistakes do happen. If you have found a mistake in this book, we would be grateful if you would report this to us. Please visit www.packtpub.com/support/errata and fill in the form.

Piracy: If you come across any illegal copies of our works in any form on the internet, we would be grateful if you would provide us with the location address or website name. Please contact us at copyright@packt.com with a link to the material.

If you are interested in becoming an author: If there is a topic that you have expertise in and you are interested in either writing or contributing to a book, please visit authors.packtpub.com.

Share Your Thoughts

Once you've read *Agile Machine Learning with DataRobot*, we'd love to hear your thoughts! Scan the QR code below to go straight to the Amazon review page for this book and share your feedback.

https://packt.link/r/1801076804

Your review is important to us and the tech community and will help us make sure we're delivering excellent quality content.

Section 1: Foundations

This section will cover some basic but critical items for the success of an ML project. Whether you are just starting or are an experienced data scientist, you will find some topics that you might not be familiar with or have skipped in the past.

This section comprises the following chapters:

- *Chapter 1, What Is DataRobot and Why You Need It*
- *Chapter 2, Machine Learning Basics*
- *Chapter 3, Understanding and Defining Business Problems*

1
What Is DataRobot and Why You Need It?

Machine learning (ML) and AI are all the rage these days, and it is clear that these technologies will play a critical role in the success and competitiveness of most organizations. This will create considerable demand for people with data science skills.

This chapter describes the current practices and processes of building and deploying ML models and some of the challenges in scaling these approaches to meet the expected demand. The chapter then describes what **DataRobot** is and how **DataRobot** addresses many of these challenges, thus allowing analysts and data scientists to quickly add value to their organizations. This chapter also helps executives understand how they can use DataRobot to efficiently scale their data science practice without the need to hire a large staff with hard-to-find skills, and how DataRobot can be leveraged to increase the effectiveness of your existing data science team. This chapter covers various components of DataRobot, how it is architected, how it integrates with other tools, and different options to set it up on-premises or in the cloud. It also describes, at a high level, various user interface components and what they signify.

By the end of this chapter, you will have learned about the core functions and architecture of DataRobot and why it is a great enabler for data analysts as well as experienced data scientists for solving the most critical challenges facing organizations as they try to extract value from data.

In this chapter, we're going to cover the following topics:

- Data science practices and processes
- Challenges associated with data science
- DataRobot architecture
- DataRobot features and how to use them
- How DataRobot addresses data science challenges

Technical requirements

This book requires that you have access to DataRobot. DataRobot is a commercial piece of software, and you will need to purchase a license for it. Most likely your organization has already purchased DataRobot licenses, and your administrator can set up your account on a DataRobot instance and provide you with the appropriate URL to access DataRobot.

A trial version is available, at the time of the writing of this book, that you can access from DataRobot's website: `https://www.datarobot.com/trial/`. Please be aware that the trial version does not provide all of the functionality of the commercial version, and what it provides may change over time.

Data science processes for generating business value

Data science is an emerging practice that has seen a lot of hype. Much of what it means is under debate and the practice is evolving rapidly. Regardless of these debates, there is no doubt that data science methods can provide business benefits if used properly. While following a process is no guarantee of success, it can certainly improve the odds of success and allow for improvement. Data science processes are inherently iterative, and it is important to not get stuck in a specific step for too long. People looking for predictable and predetermined timelines and results are bound to be disappointed. By all means, create a plan, but be ready to be nimble and agile as you proceed. A data science project is also a discovery project: you are never sure of what you will find. Your expectations or your hypotheses might turn out to be false and you might uncover interesting insights from unexpected sources.

There are many known applications of data science and new ones are being discovered every day. Some example applications are listed here:

- Predicting which customer is most likely to buy a product

- Predicting which customer will come back

- Predicting what a customer will want next

- Predicting which customer might default on a loan

- Predicting which customer is likely to have an accident

- Predicting which component of a machine might fail

- Forecasting how many items will be sold in a store

- Forecasting how many calls the call center will receive tomorrow

- Forecasting how much energy will be consumed next month

Figure 1.1 shows a high-level process that describes how a data science project might go from concept to value generation:

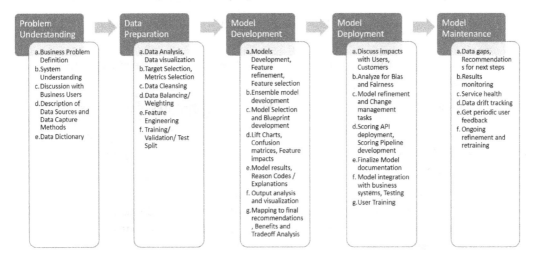

Figure 1.1 – Typical process steps with details about what happens during each step

Following these steps is critical for a successful machine learning project. Sometimes these steps get skipped due to deadlines or issues that inevitably surface during development and debugging. We will show how using DataRobot helps you avoid some of the problems and ensure that your teams are following best practices. These steps will be covered in great detail, with examples, in other chapters of this book, but let's get familiar with them at a high level.

Problem understanding

This is perhaps the most important step and also the step that is given the least attention. Most data science projects fail because this step is rushed. This is also the task where you have the least methods and tools available from the data science disciplines. This step involves the following:

- Understanding the business problem from a systemic perspective
- Understanding what it is that the end users or consumers of the model's results expect
- Understanding what the stakeholders will do with the results
- Understanding what the potential sources of data are and how the data is captured and modified before it reaches you
- Assessing whether there are any legal concerns regarding the use of data and data sources
- Developing a detailed understanding of what various features of the datasets mean

Data preparation

This step is well known in the data science community as data science teams typically spend most of their time in this step. This is a task where DataRobot's capabilities start coming into play, but not completely. There is still a lot of work that the data science or data engineering teams have to do using SQL, Python, or R. There are also many tasks in this step that require a data scientist's skill and experience (for example, feature engineering), even though DataRobot is beginning to provide capabilities in this area. For example, DataRobot provides a lot of useful data visualizations and notifications about data quality, but it is up to the analyst to make sense out of them and take appropriate actions.

This step also involves defining the expected result (such as predicting how many items will be sold next week or determining the probability of default on a loan) of the model and how the quality of results will be measured during model development, validation, and testing stages.

Model development

This step involves the development of several models using different algorithms and optimizing or tuning hyperparameters of the algorithms. Results produced by the models are then evaluated to narrow down the model list, potentially drop some of the features, and fine-tune the hyperparameters.

It is also common to look at feature effects, feature importance, and partial dependence plots to engineer additional features. Once you are satisfied with the results, you start thinking about how to turn the predictions and explanations into useable and actionable information.

Model deployment

Upon completion of model development, the model results are reviewed with users and stakeholders. This is the point at which you should carefully assess how the results will be turned into actions. What will the consequences of those actions be, and are there any unintended consequences that could emerge? This is also the time to assess any fairness or bias issues resulting from the models. Make sure to discuss any concerns with the users and business leaders.

DataRobot provides several mechanisms to rapidly deploy the models as REST APIs or executable Java objects that can be deployed anywhere in the organization's infrastructure or in the cloud. Once the model is operational as an API, the hard part of change management starts. Here you have to make sure that the organization is ready for the change associated with the new way of doing business. This is typically hard on people who are used to doing things a certain way. Communicating why this is necessary, why it is better, and how to perform new functions are important aspects that frequently get missed.

Model maintenance

Once the model is successfully deployed and operating, the focus shifts to managing the model operations and maintenance. This includes identifying data gaps and other recommendations to improve the model over time as well as refining and retraining the models as needed. Monitoring involves evaluating incoming data to see whether the data has drifted and whether the drift requires action, monitoring the health of the prediction services, and monitoring the results and accuracy of the model outputs. It is also important to periodically meet with users to understand what the model does well and where it can be improved. It is also common to sometimes employ champion and challenger models to see whether a different model is able to perform better in the production setting.

As we outlined before, although these steps are presented in a linear fashion, in practice these steps do not occur in this exact sequence and there is typically plenty of iteration before you get to the final result. ML model development is a challenging process, and we will now discuss what some of the challenges are and how to address them.

Challenges associated with data science

It is no secret that getting value from data science projects is hard, and many projects end in failure. While some of the reasons are common to any type of project, there are some unique challenges associated with data science projects. Data science is still a relatively young and immature discipline and therefore suffers from problems that any emerging discipline encounters. Data science practitioners can learn from other mature disciplines to avoid some of the mistakes that others have learned to avoid. Let's review some of the key issues that make data science projects challenging:

- **Lack of good-quality data**: This is a common refrain, but this is a problem that is not likely to go away anytime soon. The key reason is that most organizations are used to collecting data for reporting. This tends to be aggregate, success-oriented information. Data needed for building models, on the other hand, needs to be detailed and should capture all outcomes. Many organizations invest heavily in data and data warehouses in response to the need for data; the mistake they make is collecting it from the perspective of reporting rather than modeling. Hence, even after all the time and costs spent, they end up in a place where enough useable data is not available. This leads to frustration in senior leadership as to why their teams cannot make use of these large data warehouses built at enormous expense. Taking some time in developing a systemic understanding of the business can help mitigate this problem, as discussed in the following chapters.

- **Explosion of data**: Data is being generated and collected on an exponential scale. As more data is collected, the scale of the data makes it harder to be analyzed and understood through traditional reporting methods. New data also spawns new use cases that were previously not possible. The scaling of data also increases noise. This makes it increasingly difficult to extract meaningful insights with traditional methods.

- **Shortage of experienced data scientists**: This is another topic that gets a lot of press. The reason for the shortage is that it is a relatively new field where techniques and methods are still rapidly evolving. Another factor is that data science is a multi-disciplinary field that requires expertise in multiple areas, such as statistics, computer science, and business, as well as knowledge of the domain where it is to be applied. Most of the talent pool today is relatively inexperienced and therefore most data scientists have not had a chance to work on a variety of use cases with a broad range of methods and data types. Best practices are still evolving and are not in widespread use. As more and more jobs become data-driven, it will also become important for a broad range of employees to become data-savvy.

- **Immature tools and environments**: Most of the tools and environments being used are relatively immature, and that makes it difficult to efficiently build and deploy models. Most of a data scientist's time is spent wrestling with data and infrastructure issues, which limits the time spent understanding the business problem and evaluating the business and ethical implications of models. This in turn increases the odds of failure to produce lasting business value.

- **Black box models**: As the complexity of models rises, our ability to understand what they are doing goes down. This lack of transparency creates many problems and can lead to models producing nonsensical results or, at worst, dangerous results. To make matters worse, these models tend to have better accuracy on training and validation datasets. Black box models tend to be difficult to explain to stakeholders and are therefore less likely to be adopted by users.

- **Bias and fairness**: The issue of ML models being biased and unfair has been raised recently and it is a key concern for anyone looking to develop and deploy ML models. The biases can creep into the models via biased data, biased processes, or even biased decision-making using model results. The use of black box models makes this problem much harder to track and manage. Bias and fairness are hard to detect but will be increasingly important not only for an organization's reputation but also with regard to the regulatory or legal problems that they can create.

Before we discuss how to address these challenges, we need to introduce you to DataRobot because, as you might have guessed, DataRobot helps in addressing many of these challenges.

DataRobot architecture

DataRobot is one of the most well-known commercial tools for **automated ML (AutoML)**. It only seems appropriate that the technology meant to automate everything should itself benefit from automation. As you go through the data science process, you will realize that there are many tasks that are repetitive in nature and standardized enough to warrant automation. DataRobot has done an excellent job of capturing such tasks to increase the speed, scale, and efficiency of building and deploying ML models. We will cover these aspects in great detail in this book. Having said that, there are still many tasks and aspects of this process that still require decisions, actions, and tradeoffs to be done by data scientists and data analysts. We will highlight these as well. The following figure shows a high-level view of the DataRobot architecture:

Users	Data Scientists	Data Analysts	Data Engineers	Business Users	Other Systems
External Interactions	User Interface	DataRobot Client	API	Applications	
Core Functions	Data Prep (Paxata)	EDA	Modeling	MLOps	
Data Sources	Local Files	Databases	URL	HDFS	AI Catalog
Hosting Platform	On-Premises	Cloud	Hybrid Cloud	Fully Managed Cloud (SAAS)	

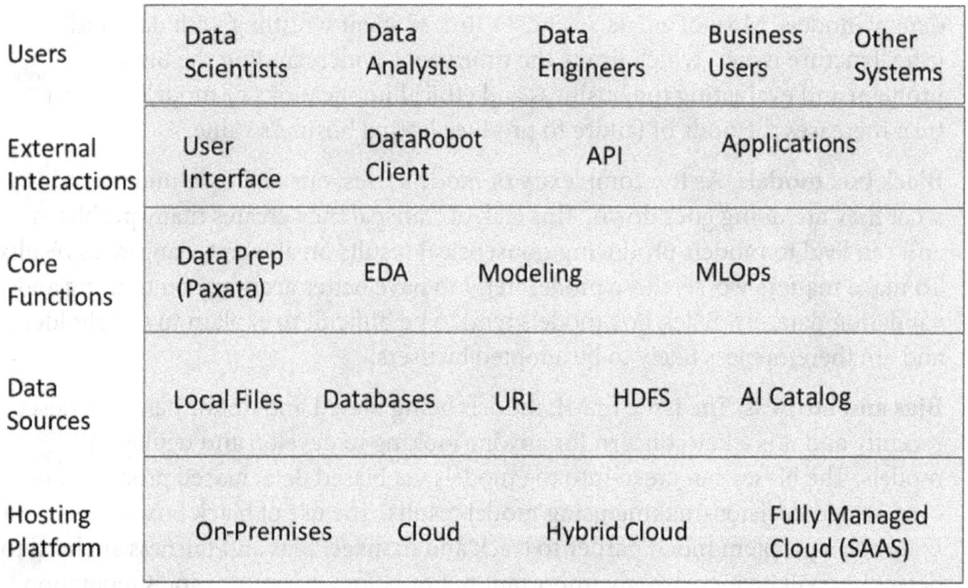

Figure 1.2 – Key components of the DataRobot architecture

The figure shows five key layers of the architecture and the corresponding components. In the following sections, we will describe each layer and how it enables a data science project.

Hosting platform

The DataRobot environment is accessed via a web browser. The environment itself can be hosted on an organization's servers, or within an organization's server instances on a cloud platform, such as AWS or DataRobot's cloud. There are pros and cons to each hosting option and which option you should choose depends on your organization's needs. Some of these are discussed at a high level in *Table 1.1*:

OPTION	PROS	CONS
Hosted on the organization's servers	Might be required for compliance reasons or due to internal policies.	Need IT resources to set up and maintain additional servers and maintain the software. Not very easy to scale up and down, so you need to provision the resources for peak loads. Downtime and effort associated with version upgrades.
Hosted on the organizaton's instances on a cloud platform or on a hybrid cloud	You control the instances, their access, and networks. Cloud instances can be easily provisioned and spun up and down based on load. Different components can live on different cloud platforms and models can be moved from one to another.	The instances are not physically under your direct control. You still need some IT resources to manage the cloud instances, set up secure networks, and manage access controls. There is still some downtime and IT resources associated with version upgrades.
Hosted on DataRobot's cloud platform	Minimal IT resource involvement as DataRobot is responsible for managing the infrastructure. You do not need to worry about IT resources for version upgrades.	The instances and data sit outside your firewalls.

Figure 1.3 – Pros and cons of various hosting options

As you can gather from this table, DataRobot offers you a lot of choices, and you can pick the option that suits your environment the best. It is important to get your IT, information security, and legal teams involved in this conversation. Let's now look at how data comes into DataRobot.

Data sources

Datasets can be brought into DataRobot via local files (csv, xlsx, and more), by connecting to a relational database, from a URL, or from **Hadoop Distributed File System (HDFS)** (if it is set up for your environment). The datasets can be brought directly into a project or can be placed into an AI catalog. The datasets in the catalog can be shared across multiple projects. DataRobot has integrations and technology alliances with several data management system providers.

Core functions

DataRobot provides a fairly comprehensive set of capabilities to support the entire ML process, either through the core product or through add-on components such as Paxata, which provides easy-to-use data preparation and **Exploratory Data Analysis (EDA)** capabilities. Discussion of Paxata is beyond the scope of this book, so we will provide details of the capabilities of the core product. DataRobot automatically performs several EDA analyses that are presented to the user for gaining insights into the datasets and catching any data quality issues that may need to be fixed.

The automated modeling functions are the most critical capability offered by DataRobot. This includes determining the algorithms to be tried on the selected problem, performing basic feature engineering, automatically building models, tuning hyperparameters, building ensemble models, and presenting results. It must be noted that DataRobot mostly supports supervised ML algorithms and time series algorithms. Although there are capabilities to perform **Natural Language Processing** (**NLP**) and image processing, these functions are not comprehensive. DataRobot has also been adding to MLOps capabilities recently by providing functions for rapidly deploying models as REST APIs, monitoring data drift and service health, and tracking model performance. DataRobot continues to add capabilities such as support for geospatial data and bias detection.

These tasks are normally done by using programming languages such as R and Python and can be fairly time-consuming. The time spent coding up data analysis, model building, output analysis, and deployment can be significant. Typically, a lot of time is also spent debugging and fixing errors and making the code robust. Depending on the size and complexity of the model, this can take anywhere from weeks to months. DataRobot can reduce this time to days. This time can in turn be used to deliver projects faster, build more robust models, and better understand the problem being solved.

External interactions

DataRobot functions can be accessed via a comprehensive user interface (which we will describe in the next section), a client library that can be used in a Python or R framework to programmatically access DataRobot capabilities via an API, and a REST API for use by external applications. DataRobot also provides the ability to create applications that can be used by business users to enable them to make data-driven decisions.

Users

While most people believe that DataRobot is for data analysts and data scientists who do not like to code, it offers significant capabilities for data scientists who can code and can significantly increase the productivity of any data science team. There is also some support for business users for some specific use cases. Other systems can integrate with DataRobot models via the API, and this can be used to add intelligence to external systems or to store predictions in external databases. Several tool integrations exist through their partners program.

Navigating and using DataRobot features

Now that you have some familiarity with the core functions, let's take a quick tour of what DataRobot looks like and how you navigate the various functions. This section will introduce DataRobot at a high level, but don't worry: we will get into details in subsequent chapters. This section is only meant to familiarize you with DataRobot functionality.

Your DataRobot administrator will provide you with the appropriate URL and a username and password to access your DataRobot instance. In my experience, Google Chrome seems to work best with DataRobot, but you can certainly try other browsers as you see fit.

> **Note**
>
> Please note that the screens and options you see depend on the products you have the license for and the privileges granted to you by your admin. For most part, it will not affect the flow of this book. Since we will be focusing on the ML development core of DataRobot, you should be able to follow along.

So, let's go ahead and launch the browser and go to your DataRobot URL. You will see a login screen as shown in the following figure:

Figure 1.4 – DataRobot login screen

Go ahead and log in using your credentials. Once you have logged in, you will be presented with a welcome screen (*Figure 1.4*) that prompts you to select what you want to do next. It is also possible that (depending on your setup) you will be directly taken to the data input screen (*Figure 1.5*):

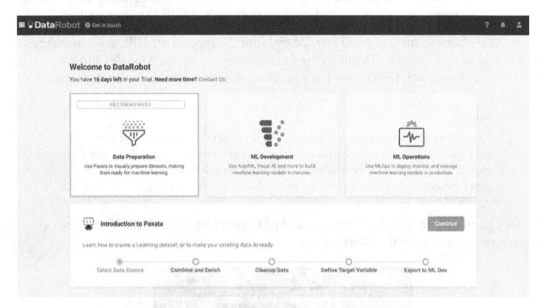

Figure 1.5 – Welcome screen

At this point, we will select the **ML Development** option and click the **Continue** button. This prompts you to provide the dataset that you wish to build models with (*Figure 1.5*):

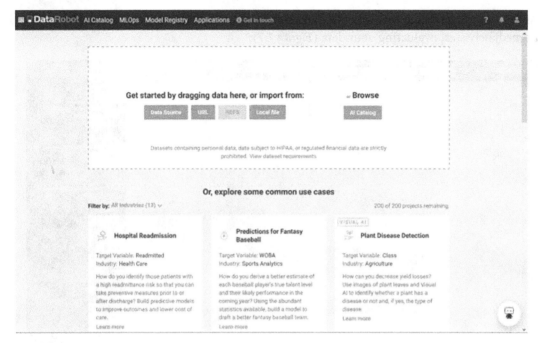

Figure 1.6 – New project/drag dataset screen

At this point, you can drag a dataset file from your local machine onto the screen (or select one of the other choices) and DataRobot will start the process of analyzing your data. You can click on the **View dataset** requirements link to see the file format options available (*Figure 1.6*). The file size requirements for your instance might be different from what you see here:

DataRobot Trial Dataset Requirements   ✕

Are you ready to experience the power of AI?
Follow these dataset requirements to put your project on the fast-track to success.

- Supported file types: .csv, .tsv, .dsv, .xls, .xlsx, .sas7bdat, .geojson, .bz2, .gz, .zip, .tar, .tgz

- Supported variable types: numeric, categorical, boolean, text, date, currency, percentage, length, and image

- Maximum dataset size: 100 MB

- Minimum rows allowed: 20

- Datasets containing personal data, data subject to HIPAA, or regulated financial data are strictly prohibited.

Got it   Learn more

Figure 1.7 – Dataset requirements

At this point, you can upload any test dataset from your local drive. DataRobot will immediately start evaluating your data (*Figure 1.7*):

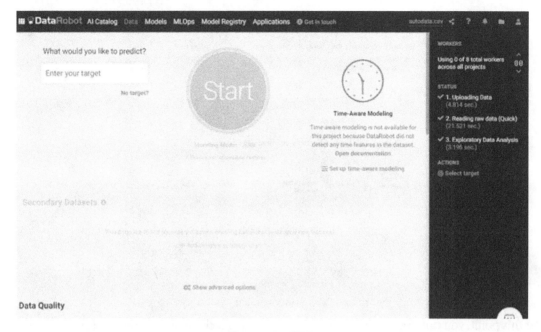

Figure 1.8 – EDA

We will cover the process of building the project and associated models in later chapters; for now, let's cover what other options we have. If you click on the **?** icon in the top right, you will see the **DOCUMENTATION** drop-down menu (*Figure 1.8*):

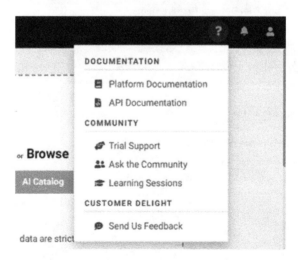

Figure 1.9 – DOCUMENTATION drop-down menu

Here you see various options to learn more about different functions, contact customer support, or interact with the DataRobot community. I highly recommend joining the community to interact with and learn from other community members. You can reach the community via `https://community.datarobot.com`. If you select **Platform Documentation** from the dropdown, you will see extensive documentation on DataRobot functions (*Figure 1.9*):

Figure 1.10 – DataRobot platform documentation

You can review the various topics at your leisure or come back to a specific topic as needed according to the task you are working on. Let's click on the **?** icon in the top right again and this time select **API Documentation** from the dropdown. You will now see the documentation for the DataRobot API (*Figure 1.10*):

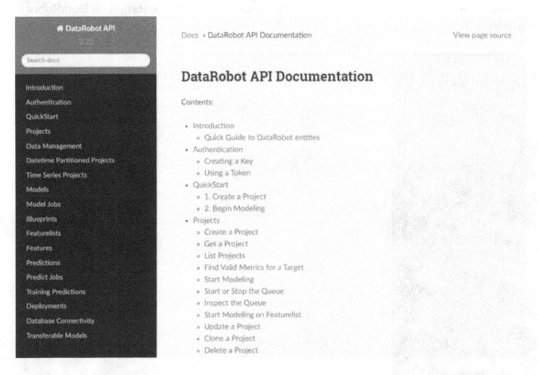

Figure 1.11 – DataRobot API Documentation

We will cover the API in the advanced topics in later chapters. If you are not familiar with programming or are relatively new to programming, you can ignore this part for now. If you are an experienced data scientist with expertise in Python or R, you can start reviewing the various functions available to you to automate your model-building tasks even further.

Let's go back to the main DataRobot page and this time select the folder icon in the top right of the page (*Figure 1.11*):

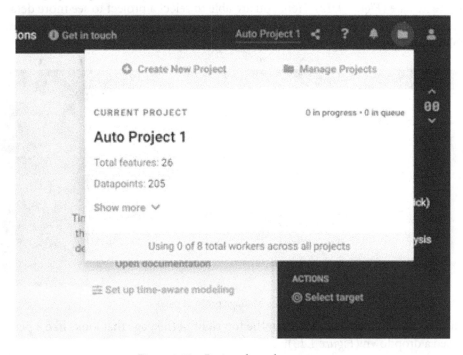

Figure 1.12 – Project drop-down menu

If you do not see the folder icon, it simply means that you do not have any projects defined. We will describe creating projects in more detail later. For now, just familiarize yourself with different options and what they look like. Here you will see options to create a new project or manage existing projects. In here, you will also see some details about the currently active project as well as a list of recent projects.

The **Create New Project** option takes you back to the new project page that we saw before in *Figure 1.5*. If you select the **Manage Projects** menu, it will show all of your projects listed by create date (*Figure 1.12*). Here you are able to select a project to see more details, clone a project, share the project with other users, or delete a project as needed, as shown in the following figure:

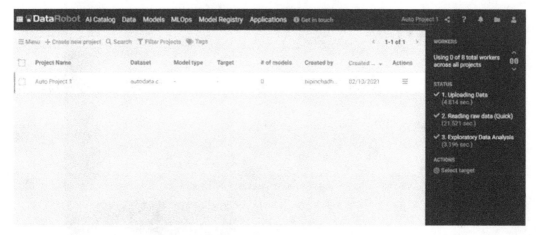

Figure 1.13 – Manage projects page

If you click on the very last menu item in the top right of the page that looks like a person, you will see a dropdown (*Figure 1.13*):

Figure 1.14 – User account management dropdown

From here you can manage your profile and adjust your account settings. If you have admin privileges, you can view and manage other users and groups. You can also sign out of DataRobot if needed.

If you select the **Profile** menu, you will see details of your account (*Figure 1.14*):

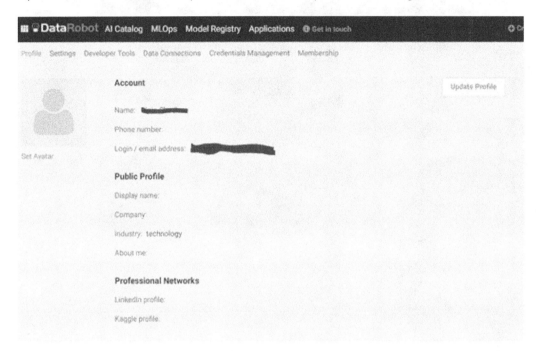

Figure 1.15 – User profile page

Here you can update some of your information. You will also see some new menu choices on the second menu row at the top. This allows you to change settings or access some developer options, and so on. If you select the **Settings** menu, you will see the following (*Figure 1.15*):

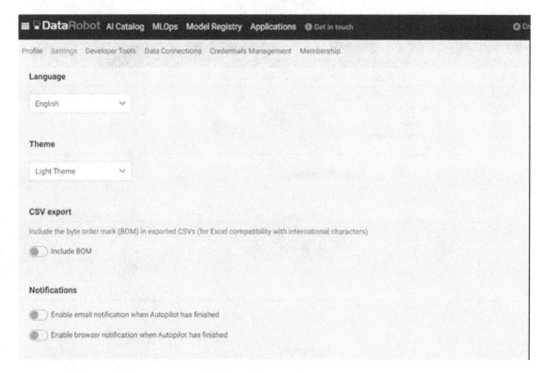

Figure 1.16 – User Settings

On this page, you can change your password, set up two-factor authentication, change the theme, and set up notifications (you will see different options available to you based on how your account was set up by your administrator).

If you select **Developer Tools**, you will see the following (*Figure 1.16*):

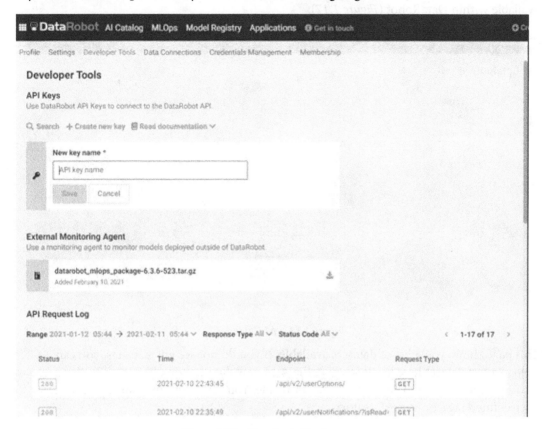

Figure 1.17 – Developer Tools screen

Here you can create an API key associated with your account. This key is useful for authentication if you will be using the DataRobot API. You can also download the API package to set up a portable prediction server to deploy models within your organization's infrastructure.

If you click on the **AI Catalog** menu at the top, you will see a catalog of shareable datasets available within DataRobot (*Figure 1.17*):

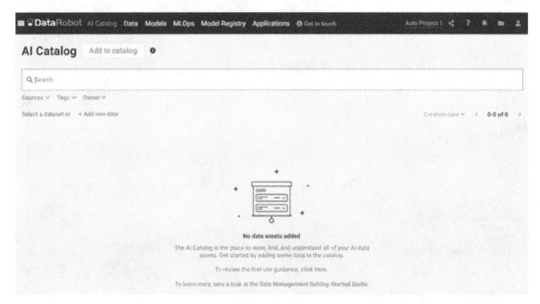

Figure 1.18 – AI Catalog

This page shows you a list of datasets available. If you do not see any datasets, you can upload a test dataset here by clicking on the **Add new data** button (*Figure 1.18*). You can also click on a dataset to explore the data available. You can search and sort by sources, user-defined tags, or owner/creator:

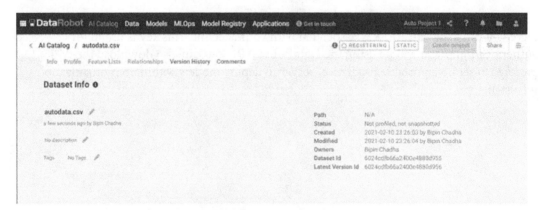

Figure 1.19 – Dataset information page

Normally a dataset is only available within a project. If you want to share datasets across projects or iterations of projects, you can create the dataset within this catalog. This allows you to share these datasets across projects and users. The datasets can be static, or they can be dynamically created using a SQL query as needed. Datasets can also be modified or blended via Spark SQL if you need data from multiple tables or sources for a project.

If you click on the **Profile** button, you will see profile-level information about the dataset (*Figure 1.19*). This information is automatically compiled for you. We will describe these capabilities and how to use them in more detail later:

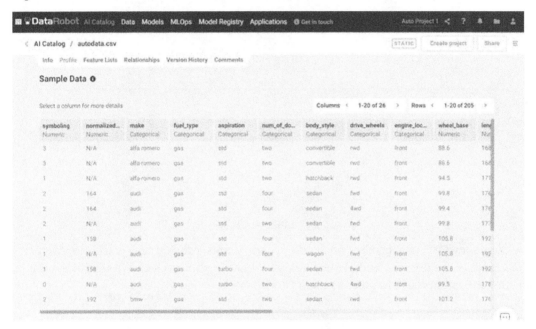

Figure 1.20 – Dataset information page

This page shows details of the dataset that is part of the project that is active at that time. This page is one of the key components of the DataRobot capability. The page shows summary information as well as any data quality issues that DataRobot has detected. Below that, it shows summaries of data features as well as a feature's importance relative to the target feature. We will cover these capabilities in more detail in subsequent chapters.

Let's now click on the **Data** menu at the top left of the page. This page (*Figure 1.20*) shows a detailed analysis of the dataset for your currently active project:

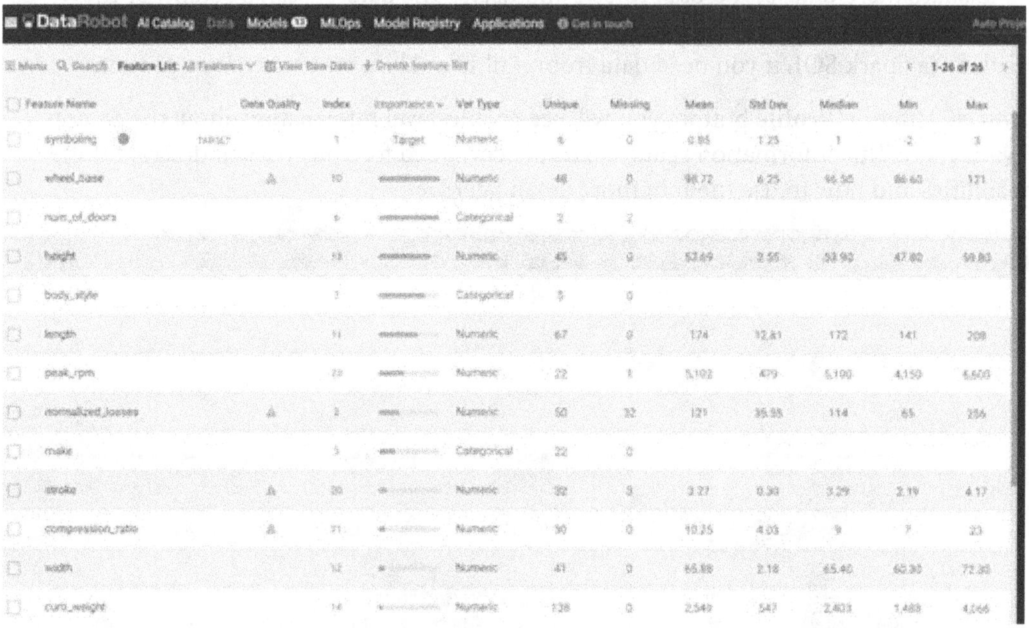

Figure 1.21 – Project data page

This page shows the results of the analysis of your datasets, provides any warnings, relative importance of the features, and the feature lists for use in your project. We will review the functionality of this page in great detail in later chapters.

Let's now click on the **Models** menu item at the top. This shows the model leaderboard for the active project (*Figure 1.21*):

Figure 1.22 – Model leaderboard

This is another critical page where you will spend a lot of your time during the modeling process. Here you can see the top-performing models that DataRobot has built and their performance metrics for validation, cross-validation, and holdout samples. You can drill down into the details of any selected model. It is important to note that DataRobot mostly works with supervised learning problems; currently, it does not have support for unsupervised learning (except for some anomaly detection) or reinforcement learning. Also, support for NLP and image processing problems is limited. Similarly, there are situations where either due to data limitations or extreme scales, you will find that the automation adds a level of overhead that makes it impractical to use DataRobot. If your project requires advanced capabilities in these areas, you will need to work in Python or R directly. More on this in subsequent chapters.

Let's now move to the next menu item, **MLOps**. When you click on **MLOps,** you will see the screen shown in *Figure 1.22*:

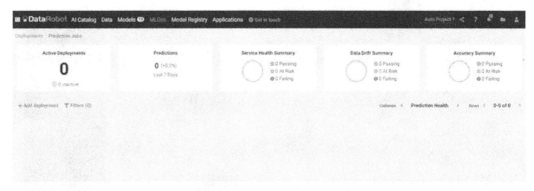

Figure 1.23 – MLOps page

The **MLOps** page shows you your active deployments and their health. You can set up alerts relating to data drift or model accuracy as needed for your use cases.

The next menu item is **Model Registry**. Now, **Model registry** is the mechanism by which you can bring externally developed models into DataRobot. This capability is an add-on that your organization may or may not have purchased. This aspect is an advanced topic that is beyond the scope of this book.

Let's click on the next menu item, **Applications**. You will now see what's shown in *Figure 1.23*:

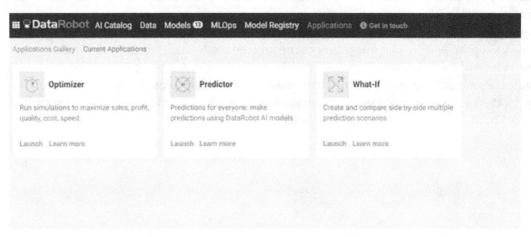

Figure 1.24 – Applications page

Applications is a relatively new functionality in DataRobot that is meant to allow business users to easily access model results without needing to get DataRobot user licenses.

This concludes our quick tour of what DataRobot is and what it looks like. We will revisit many of these components in great detail and see examples of how these are used to take a data science project from start to finish.

Addressing data science challenges with DataRobot

Now that you know what DataRobot offers, let's revisit the data science process and challenges to see how DataRobot helps in addressing these challenges and why this is a valuable tool in your toolkit.

Lack of good-quality data

While DataRobot cannot do much to address this challenge, it does offer some capabilities to handle data with quality problems:

- Automatically highlights data quality problems.
- Automated EDA and data visualization expose issues that could be missed.
- Handles and imputes missing values.
- Detection of data drift.

Explosion of data

While it is unlikely that the increase in the volume and variety will slow down any time soon, DataRobot offers several capabilities to address these challenges:

- Support for SparkSQL enables the efficient pre-processing of large datasets.
- Automatically handles categorical data encodings and selects appropriate model blueprints.
- Automatically handles geospatial features, text features, and image features.

Shortage of experienced data scientists

This is a key challenge for most organizations and data science teams, and DataRobot is well positioned to address this challenge:

- Provides capabilities that cover most of the data science process steps.

- Significant automation of several routine tasks by providing pre-built blueprints encoded with best practices.

- Experienced data scientists can build and deploy models much faster.

- Data analysts or data scientists who are not very comfortable coding can utilize DataRobot capabilities without having to write a lot of code.

- Experienced data scientists who are comfortable with coding can utilize the APIs to automatically build and deploy an order of magnitude more models than otherwise feasible without the support of other data engineering or IT staff.

- Even experienced data scientists do not know all the possible algorithms and typically do not have the time to try out many of the combinations and build analysis visualizations and explanations for all models. DataRobot takes care of many of these tasks for them, enabling them to focus more time on understanding the problem and analyzing results.

Immature tools and environments

This is a key barrier to the productivity and effectiveness of any data science organization. DataRobot clearly addresses this key challenge by offering the following:

- Ease of deployment of any model as a REST API.

- Ease of use in developing multiple competing models and selecting the best ones without worrying about the underlying infrastructure, installation of compatible versions, and without coding and debugging. These tasks can take up a lot of time that would be better spent on understanding and solving the business problem.

- DataRobot encodes many of the best practices into their development process so as to prevent mistakes. DataRobot automatically takes care of many small details that can be overlooked even by experienced data scientists, leading to flawed models or rework.

- DataRobot provides automated documentation of models and modeling steps that could otherwise be glossed over or forgotten. This becomes valuable at a later time when a data scientist has to revisit an old model built by them or someone else.

Black box models

This is a key challenge that DataRobot has done extensive work on to provide methods to help make models more explainable, such as the following:

- Automated generation of feature importance (using Shapley values and other methods) and partial dependence plots for models

- Automated generation of explanations for specific predictions

- Automated generation of simpler models that could be used to explain the complex models

- Ability to create models that inherently more explainable such as **Generalized Additive Models (GAMs)**

Bias and fairness

Recently, DataRobot has added capabilities to help detect bias and fairness issues in models. This is no guarantee of a complete lack of bias, but it's a good starting point to ensure positive movement in this direction. Some of the capabilities added are listed here:

- Specify protected features that need to be checked for bias.

- Specify bias metrics that you want to use to check for fairness.

- Evaluate your models using metrics for protected features.

- Use of model explanations to investigate whether there is potential for unfairness.

While many people believe that with these automated tools, you no longer need data scientists, nothing could be further from the truth. It is, however, obvious that such tools will make data science teams a lot more valuable to their organizations by unlocking more value faster and by making these organizations more competitive. It is therefore likely that tools such as DataRobot will become increasingly commonplace and see widespread use.

Summary

Most data scientists today are bogged down in the implementation details or are implementing suboptimal algorithms. This leaves them with less time to understand the problem and to search for optimal algorithms or their hyperparameters. This book will show you how to take your game to the next level and let the software do the repetitive work.

In this chapter, we covered what a typical data science process is and how DataRobot supports this process. We discussed steps in the process where DataRobot offers a lot of capability and we also highlighted areas where a data scientist's expertise and domain understanding is critical (areas such as problem understanding and analyzing the impacts of deploying a model on the overall system). This highlights an important point in that success comes from the combination of skilled data scientists and analysts and appropriate tools (such as DataRobot). By themselves, they cannot be as effective as the combination. DataRobot enables relatively new data scientists to quickly develop and deploy robust models. At the same time, experienced data scientists can use DataRobot to rapidly explore and build a broader range of models than they would be able to build on their own.

We covered some of the key data science challenges and how DataRobot helps you overcome some of the specific challenges. This should help guide leaders on how to craft the right combination of data scientists and the tools and infrastructure they need. We also covered the DataRobot architecture, its components, and what DataRobot looks like. You got a taste of what you will see when you start using it and where to go to find specific functions and help.

Hopefully, this chapter has shown you why DataRobot could be an important tool in your toolbox regardless of your experience or how comfortable you are with coding. In the following chapters, we will use hands-on examples to show how to use DataRobot in detail and how to move your projects into a higher gear. But before we do that, we need to cover some ML basics in the next chapter.

2
Machine Learning Basics

This chapter covers some basic concepts of machine learning that will be used and referenced in this book. This is the bare minimum you need to know in order to use DataRobot effectively. Experienced data scientists can safely skip this chapter. It is not the intention of this chapter to give you a comprehensive understanding of statistics or machine learning, but just a refresher of some key ideas and concepts. Also, the focus is on practical aspects of what you need to know in order to understand the core ideas without going into too much detail. It might be tempting to jump in and let DataRobot automatically build the models, but doing that without a basic understanding could backfire. If you are leading a data science team, please make sure that you have experienced data scientists in your teams who are mentoring others and that there are other governance processes in place.

Some of these concepts will come up again during the hands-on examples, but we are covering many concepts here that might not come up during a specific example, but might come up in relation to your project at some point. The topics listed here can be used as a guide to determine some of the basic knowledge that you require in order to start using powerful tools such as DataRobot.

By the end of this chapter, you will have learned some of the core concepts you need to know to use DataRobot effectively. In this chapter, we're going to cover the following main topics:

- Data preparation
- Data visualization
- Machine learning algorithms
- Performance metrics
- Understanding the results

Data preparation

Before an algorithm can be applied to a dataset, the dataset needs to fit a certain pattern. The dataset also needs to be free of errors. Certain methods and techniques are used to ensure that the dataset is ready for the algorithms, and this will be the focus of this section.

Supervised learning dataset

Since DataRobot mostly works with supervised learning problems, we will only focus on datasets for supervised machine learning (other types will be covered in a later section). In a supervised machine learning problem, we provide all the answers as part of the dataset. Imagine a table of data where each row represents a set of clues with their corresponding answers (*Figure 2.1*):

Clue1 (# of legs)	Clue2 (# of eyes)	Clue3 (# of hands)	Clue4 (furry)	Answer
4	2	0	Y	Dog
2	2	2	N	Person
3	2	0		Dog
2	2	2	N	Human

Figure 2.1 – Supervised learning dataset

This dataset is made up of columns that contain clues (these are called **features**), and there is a column with the answers (this is called **target**). Given a dataset that looks like this, the algorithm learns how to produce the right answer given a set of clues. No matter what form your data is in, your task is to first transform it to make it look like the table in *Figure 2.1*. Note that the clues that you have might be spread across multiple databases or Excel files. You will have to compile all of that information into one table. If the datasets you have are complex, you will need to use languages such as SQL, tools such as **Python Pandas**, or **Excel**, or tools such as **Paxata**.

Time series datasets

Time series or forecasting problems have time as a key component of their datasets. They are similar to the supervised learning datasets, with slight differences, as shown in *Figure 2.2*:

Date	Clue2 (Holiday)	Clue3 (Product Announcement)	Answer (Units Sold)
1	0	0	34
2	0	1	56
3	1	0	57
4	0	0	59

Figure 2.2 – Time series dataset

You need to make sure that your time series datasets appear as shown in the preceding diagram. It should have a date or time-based column, and a column with the series values you are trying to forecast, and a set of clues as needed. You can also add columns that help to categorize different series, if there are multiple time series that you need to forecast. For example, you might be interested in forecasting units sold for dates 5 and 6. If your data is in some other form, it needs to be transformed to look like the preceding diagram.

Data cleansing

The data that comes to you will typically have errors in it. For example, you might have text in a field that is supposed to contain numbers. You might see a price column where the values may contain a $ sign on occasion, but no sign at other times. DataRobot can catch some of these, but there are times when an automated tool will not catch these, so you need to look and analyze the dataset carefully. It is useful to sometimes upload your data to DataRobot to see what it finds, and then use its analysis to determine the next steps. Some of this cleansing will need to be performed outside DataRobot, so be prepared to iterate a few times to get the data set up correctly. Common issues to watch out for include the following:

- Wrong data type in a column
- Mixed data types in a column
- Spaces or other characters in numeric columns that make them look like text
- Synonyms or misspelled words
- Dates encoded as strings
- Dates with differing formats

Data normalization and standardization

When different data features have varying scales and ranges, it becomes harder to compare their impacts on the target values. Also, many algorithms have difficulty in dealing with different scales of values, sometimes leading to stability issues. One method for avoiding these problems is to normalize (not to be confused with database normal forms) or standardize the values.

In normalization (also known as scaling), you scale the values such that they range from 0 to 1:

$$X_{normalized} = (X - X_{min}) / (X_{max} - X_{min})$$

Standardization, on the other hand, centers the data such that the mean becomes zero and scales it such that the standard deviation becomes 1. This is also known as `z-scoring` the data:

$$X_{standardized} = (X - X_{mean}) / X_{SD}$$

Here, X_{mean} is the mean of all X values, and X_{SD} is the standard deviation of X values.

In general, you will not need to worry about this because DataRobot automatically does this for the datasets as required.

Outliers

Outliers are values that seem to be out of place compared to the rest of the dataset. These values can be very large or very small. In general, values that are more than three standard deviations from the mean are considered outliers, but this only applies to features where values are expected to be normally distributed. Outliers typically come from data quality issues or some unusual situations that are not considered relevant enough to be trained on. The data points deemed to be outliers are typically removed from the dataset to prevent them from overpowering your models. The rules of thumb are only for highlighting the candidates. You will have to use your judgment to determine whether any values are outliers and whether they need to be removed. Once again, DataRobot will highlight potential outliers, but you will have to review those data points and determine whether to remove them or leave them in.

Missing values

This is a very common problem in datasets. Your dataset may contain many missing values, marked as **NULL** or **NaN**. In some cases, you will see a **?**, or you might see an unusual value, such as **-999**, that an organization might be using to represent a missing or unknown value. How you choose to handle such values depends a lot on the problem you are trying to solve and what the dataset represents. Many times, you might choose to remove the row of data that contains a missing value. Sometimes, that is not possible because you might not have enough data, and removing such rows might lead to the removal of a significant portion of your dataset. Sometimes, you will see a large number of values in a feature (or column) that might be missing. In those situations, you might want to remove that feature from the dataset.

Another possible way of dealing with this situation is to fill the missing values with a reasonable guess. This could take the form of a zero value, or the mean value for that feature, or a median value of that feature. For categorical data, missing values are typically treated as their own separate category.

More sophisticated methods use the k-nearest neighbor algorithm to compute missing values based on other similar data points. No one answer will be appropriate every time, so you will need to use your judgment and understanding of the problem to make a decision. One final option is to leave it as it is and let DataRobot figure out how to deal with the situation. DataRobot has many imputation strategies as well as algorithms to handle missing values. But you have to be careful, as that might not always lead to the best solution. Talk to an experienced data scientist and use your understanding of the business problem to plot a course of action.

Category encoding

In many problems, you have to transform your features into numeric values. This is because many algorithms cannot handle categorical data. There are many ways to encode categorical values and DataRobot has many of these methods built in. Some of these techniques are one-hot encoding, leave one out encoding, and target encoding. We will not get into the details, as normally you would let DataRobot handle this for you, but there might be cases where you will want to encode it yourself in a specific way due to your understanding of the business problem. This feature of DataRobot is a great time saver and typically works very well for most problems.

Consolidate categories

Sometimes, you have categorical data that contains a large number of categories. Although there are methods for dealing with large category counts (as discussed in the preceding section), many times, it is advisable to consolidate the categories. For example, you might have many categories that contain very few data points, but are very similar to one another. In this case, you can combine them into a single category. In other cases, it might just be that someone used a different spelling, a synonym, or an abbreviation. In such cases, it is better to combine them into a single category as well. Sometimes, you might want to split up a numerical feature into bins that have a business meaning for your users or stakeholders. This is an example of data preparation that you will need to do on your own based on your understanding of the problem. You should do this prior to uploading the data into DataRobot.

Target leakage

Sometimes, the dataset contains features that are derived from the target itself. These are not known in advance or are not known at the time of prediction. Inadvertently using these features to build a model causes problems downstream. This issue is called target leakage. The dataset should be inspected carefully and such features should be removed from the training features. DataRobot will also analyze the features automatically and try to flag any features that might lead to target leakage.

Term-document matrix

Your dataset may contain features that contain text or notes. These notes frequently contain important information that is useful for making decisions. Many of the algorithms, however, cannot make use of this text directly. This text has to be parsed into numeric values for it to become useful to modeling algorithms. There are several methods for doing that, with the most common one being the term-document matrices. Document here refers to a single text or notes entry. Each of these documents can be parsed to split it up into terms. Now you can count how many times a term showed up in a document. This result can be stored in a matrix called a **Term Frequency** (**TF**) matrix. Some of this information can also be visualized in word clouds. DataRobot will automatically build these word clouds for you. While TF is useful, it can be limiting because some terms might be very common in all the documents, hence they are not very useful in distinguishing between them. This leads to another idea, whereby perhaps we should look for terms that are somewhat unique to a document. This concept of giving more weight to a term that is present in some documents only is called **Inverse Document Frequency** (**IDF**). The combination of a term showing up multiple times in a document (TF) and it being somewhat rare (IDF) is called **TFIDF**. TFIDF is something that DataRobot will compute automatically for you and gets applied to features that contain text.

Data transformations

While DataRobot will do many data transformations for you (and it keeps adding more all the time), there are many transformations that will impact your model but that DataRobot will not be able to catch. You will have to do these on your own. Examples of these are mathematical transformations such as log, square, square root, absolute values, and differences. Some of the simple ones can be set up inside DataRobot, but for more complex ones, you will have to perform the operations outside of DataRobot or in tools such as Paxata. Sometimes, you will do a transformation to linearize your problem or to deal with features that have long-tailed data. Some of the transformations that DataRobot does automatically are as follows:

- Computing aggregates such as counts, min, max, average, median, most frequent, and entropy

- An extensive list of time-based features, such as change over time, max over time, and averages over time

- Some text extraction features, such as word counts, extracted tokens, and term-document matrices

- Geospatial features from geospatial data

We will discuss this topic again in more detail in *Chapter 4, Preparing Data for DataRobot.*

Collinearity checks

In any given dataset, there will be features that are highly correlated to other features. In essence, they carry the same information as some other features. It is generally desirable to remove such features that are highly duplicative of some other features in the dataset. DataRobot performs these checks automatically for you and will flag these collinear features. This is especially critical for linear models, but some of the newer methods can deal with this issue better. What thresholds to use varies based on the modeling algorithms and your business problem. It is fairly easy in DataRobot to remove these features from your feature sets to be used for modeling.

DataRobot also produces a correlation matrix that shows how the different features are correlated to one another. This helps identify collinear features as well as key candidate features to be used in the model. You can gain a lot of insight into your data and the problem by analyzing the correlation matrix. In *Chapter 5, Exploratory Data Analysis with DataRobot*, we will discuss examples of how this is done.

Data partitioning

Before you start building the models, you need to partition your dataset into three parts. These parts are called training, validation, and holdout. These three parts are used for different purposes during the model building process. It is common to split 10-20% of the dataset into the holdout set. The remaining portion is split up further, with 70-80% going to training and 20-30% going to the validation set. This splitting is done to make sure that the models are not overfitted and that the expected results in deployment are in line with results seen during model building.

Only the training dataset is used to train the model. The validation set is designed to tune the algorithms in order to optimize the results by performing multiple cross-validation tests. Finally, the holdout set is used after the models are built to test the model on data that it has never seen before. If the results on the holdout set are acceptable, then the model can be considered for deployment.

DataRobot automates most of this process, but it does allow the user to customize the split percentages, as well as how the partitioning should be done. It also performs a similar function for time series or forecasting problems by automatically splitting the data for time-based backtests.

Data visualization

One of the most important tasks a data analyst or data scientist needs to do is to understand the dataset. Data visualization is key to this understanding. DataRobot provides various ways to visualize the datasets to help you understand the dataset. These visualizations are built automatically for you so that you can spend your time analyzing them instead of preparing them. Let's look at what these are and how to use them.

When you go to the data page (*Figure 1.20*) for your project, you will see high-level profile information for your dataset. Inspect this information carefully to understand your dataset in totality. If you click on the **Feature Association** menu (top left), you will see how the features are related to one another (*Figure 2.3*):

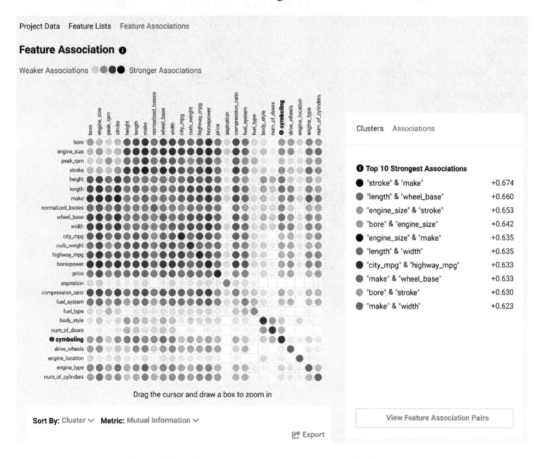

Figure 2.3 – Feature associations using mutual information

This diagram shows the interrelationships using the mutual information metric. **Mutual Information** (**MI**) uses information theory to determine the amount of information you obtain about one feature from the other feature. The benefit of using MI compared to the Pearson correlation coefficient is that it can be used for any type of feature. The value goes from 0 (the two features are independent) to 1 (they carry the same information). This is useful in determining which features will be good candidates for the model and which features will not provide any useful information or are redundant. This view is extremely important to understand and use before model building starts, even though DataRobot automatically uses this information to make modeling decisions.

There is another metric that is also used in a similar capacity. If you click on the metric dropdown at the bottom of the preceding screenshot, you can select the other metric called **Cramer's V**. Once you select Cramer's V, you will see a similar graphical view (*Figure 2.4*):

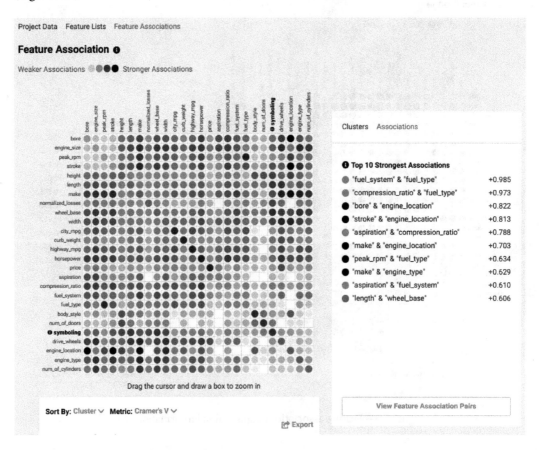

Figure 2.4 – Feature associations using Cramer's V

Cramer's V is an alternative metric to MI, and it is used similarly. Its value also ranges from 0 (no relationship) to 1 (the features are highly correlated). Cramer's V is often used with categorical variables as an alternative to the Pearson correlation coefficient.

Notice that DataRobot automatically found clusters of interrelated features. Each cluster is color-coded in a different color, and the features are sorted by clusters in *Figure 2.4*. You can zoom into specific clusters to inspect them further. This is an important feature of the DataRobot environment as very few data scientists know about this idea or make use of it. The clusters are important because they highlight groups of interrelated features. These complex interdependencies are typically very important for understanding the business problem. Normally, the only people who know about these complex interdependencies are people with a lot of domain experience. Most others will not even be aware of these complexities. If you are new to a domain, then understanding these will give you an equivalent of multiple years of experience. Study these carefully, discuss them with your business experts to fully understand what they are trying to highlight, and then use these insights to improve your models as well as your business processes.

Also, note that DataRobot provides a list of the top 10 strongest associations. It is important to note these associations and spend some time thinking about what they mean for your problem. Are these consistent with what you know about your domain, or are there some surprises? It is the surprises that often result in key insights that could prove to be valuable insights for your business. In the following list, you see a **View Feature Association Pairs** button. If you click on that button, you will see *Figure 2.5*:

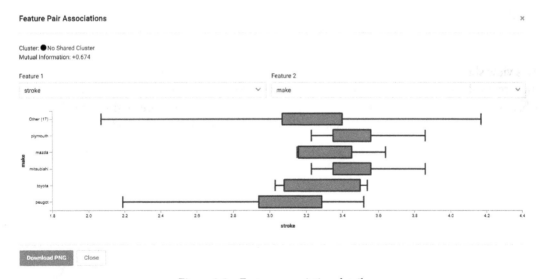

Figure 2.5 – Feature association details

This graphic shows the relationship between two selected features in detail. In this example, one feature is categorical while the other is numeric. The diagram shows how the two are related and could provide additional insights into the problem. Be sure to investigate the relationships, especially the ones that might be counterintuitive.

Now you can click on the specific features to see how they are distributed (*Figure 2.6*):

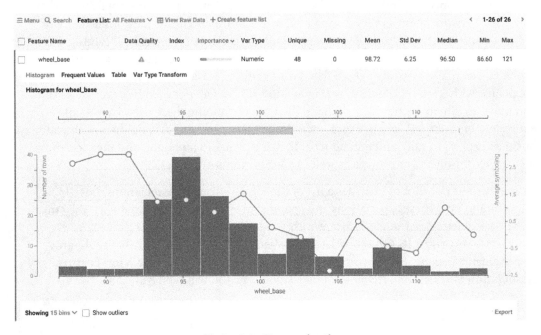

Figure 2.6 – Feature details

This view shows a histogram of how the values are distributed and how they are related to the target values. Key things to focus on are ranges where you do not have enough data and where you have non-linearities. These could give you ideas about feature engineering. These are also areas where you ask the question why does the system exhibit this behavior?

With this background work done, you are now ready to dive into modeling algorithms.

Machine learning algorithms

There are now hundreds of machine learning algorithms available to be used for a machine learning project, and more are being invented every day. DataRobot supports a wide array of open source machine learning algorithms, including several deep learning algorithms – Prophet, SparkML-based algorithms, and H2O algorithms. Let's now take a look at what types of algorithms exist and what they are used for (*Figure 2.7*):

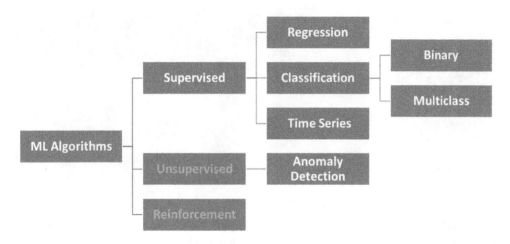

Figure 2.7 – Machine learning algorithms

Our focus will mostly be on the algorithm types that DataRobot supports. These algorithm types are described in the following sub-sections.

Supervised learning

Supervised learning algorithms are used when you can provide an answer (also called a label) as part of the training dataset. For supervised learning, you have to assign a feature of your dataset to be the answer, and the algorithm tries to learn to predict the answer by seeing multiple examples and learning from these examples. See *Figure 2.8* for the different types of answers:

Clue1 (# of legs)	Clue2 (# of eyes)	Clue3 (# of hands)	Clue4 (furry)	Clue5 (Gender)	Answer (Regression)	Answer (Binary Class)	Answer (Multi Class)
4	2	0	Y	M	1.4	Dog	A
2	2	2	N	F	2.3	Person	B
3	2	0		M	9.8	Dog	C
2	2	2	N		1.6	Person	D

Figure 2.8 – Targets for supervised learning algorithms

DataRobot functionality is primarily focused on supervised learning algorithms. Included in the set are deep learning algorithms as well as big data algorithms from SparkML and H2O. DataRobot has built-in best practices to select the best-suited algorithms for your problem and dataset. There are four major types of supervised learning problems:

Regression

Regression problems are the ones where the answer (target) takes a numeric form (see *Figure 2.8*). Regression models try to fit a curve such that the error between the prediction and the actual value is minimized for the entire training dataset. Sometimes, even a classification problem can be set up as a numeric regression problem. In such cases, the answer is a number that can then be turned into a bin by using thresholds. Logistic regression is one such method that produces a value between zero and one. You can mark all answers below a certain threshold to be zero, and all above as ones. There are linear as well as non-linear regression algorithms that are used based on the problem. The models are assessed based on how well the regression line matches the data. Typical metrics used are **RMSE, MAPE, LogLoss**, and **Rsquared**. Typical algorithms used are **XGBoost, Elastic Net, Random Forest**, and **GA2M**.

Binary classification

Binary classification problems have answers that can only take two distinct values (called classes). These could be in the form of 0 or 1, Yes or No, and so on. Please refer to *Figure 2.8* for an example of the target feature for binary classification. A typical issue that you commonly face is the problem of class imbalance. This happens when most of the dataset is biased toward one class. These are typically addressed by downsampling the overrepresented class when sufficient training data is present. When this is not possible, you can try oversampling the underrepresented class or use other methods. None of these methods is perfect, and sometimes you have to try different approaches to see what works best. DataRobot provides mechanisms to specify downsampling if needed. Some of the algorithms that are commonly used for binary classification are **logistic regression**, **k-nearest neighbors**, **tree-based algorithms**, **SVM**, and **Naïve Bayes**. In the case of classification problems, it is best to avoid using accuracy as a metric to assess results. The results are often shown in the form of a confusion matrix (described later in this chapter). DataRobot will automatically select an appropriate metric to use in such cases.

Multiclass classification

Multiclass classification problems are the ones where you are trying to predict more than two classes or categories. For a simple example of what the target might look like, see *Figure 2.8*. Multiclass capability was added recently and many of the DataRobot features might not work with such problems. Since downsampling is not available, you might want to adjust your sampling prior to uploading your dataset into DataRobot. Also, note that you can frequently collapse your problem into a binary classification problem by collapsing the classes into two classes. That may or may not work for your use case, but it is an option if required. Also, not all algorithms are appropriate for multiclass problems. DataRobot will automatically select the appropriate algorithms to build the models for multiclass problems. Typical metrics to use are AUC, LogLoss, or Balanced Accuracy. The results are often shown in the form of a confusion matrix (described later in this chapter). Typical algorithms used are XGBoost, Random Forest, and TensorFlow.

Time series/forecasting

Time series or forecasting models are also referred to as time-aware models in DataRobot. In these problems, you have data that is changing over time and you are interested in predicting/forecasting a target value in the future (*Figure 2.2*). DataRobot not only supports the usual algorithms for time series such as ARIMA, but can also adapt these problems to machine learning regression problems and then apply algorithms such as XGBoost to solve them. These problems require that the series should be transformed into stationary series and require extensive feature engineering to create time-based features. The problems also require that you take into account important events in the past that may repeat (such as holidays or major shopping days). Time series models also require special ways of handling validation and testing via a method called backtests:

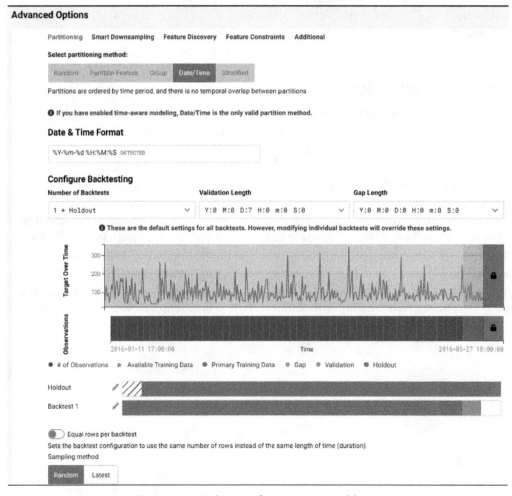

Figure 2.9 – Backtesting for time series problems

In backtesting, models are built using past data, and then tested using holdout data that is newer and has never been seen by the model. This time-based slicing of holdout data is also referred to as out-of-time validation. DataRobot automates many of these tasks for you, as we will see in more detail later.

Algorithms

Let's review some of the main algorithms used in DataRobot. Here, we only provide a high-level overview of these algorithms These algorithms can be tuned for a given problem by changing their hyperparameters. For a more detailed understanding of any specific algorithm, you can refer to a machine learning book or the DataRobot documentation. Some of the important algorithms are as follows:

- **Random Forest**. A random forest model is built by creating multiple decision tree models and then uses the mean of the output. This is done by creating bootstrap samples of the training data and building decision trees (*Figure 2.10*) on these samples:

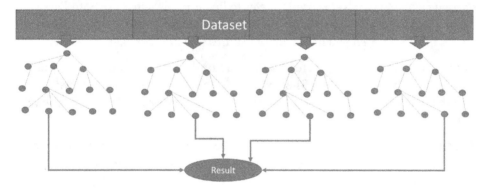

Figure 2.10 – Random forest

Random forest models handle missing data and non-linearities and have proven to work great in many situations. A random forest model can be used for regression as well as classification problems:

- **XGBoost**: Also known as **eXtreme** gradient boosted trees, are decision tree-based algorithms that have become very popular because they tend to produce very effective predictions and can handle missing values. They can handle non-linear problems and interactions between features. XGBoost builds upon random forest models by creating a random forest and then creating trees on the residuals of the previous trees. This way, every new set of trees is able to produce a better result. XGBoost can be used for regression as well as classification problems.

- **Rulefit**: Rulefit models are ensembles of simple rules. You can think of these rules as being chained together like a decision tree. Rulefit models are much easier to understand as most people can relate to a combination of rules being applied to solve a problem. DataRobot typically builds this model to help you understand a problem and provide insights. You can go to the insights section of your **Models** tab and see the insights generated from a Rulefit model and how effective a given rule is for the problem. They can be used for classification as well as regression problems.

- **ElasticNet, Ridge regressor, Lasso regressor**: These models use regularization to make sure that the models are not overfitting and are not unnecessarily complex. Regularization is done by adding a penalty for adding more features, which in turn forces the models to either drop some features or reduce their relative impact. Lasso regressor (also known as **L1 regressor**) uses penalty weights that are the absolute values of the coefficients. The effect of using Lasso is that it tries to reduce the coefficients to zero, thereby selecting important features and removing the ones that do not contribute much. Ridge regressor (also known as **L2 regressor**) uses penalty weights that are squared coefficients. The impact of this is to reduce the magnitude of coefficients. **ElasticNet** is used to refer to linear models that use both Lasso and Ridge regularization to produce models that are simpler as well as regularized. This comes in handy when you have a lot of features that are correlated with each other.

- **Logistic Regression**: Logistic regression is a non-linear regression model that is used for binary classification. The output is in the form of a probability with a value ranging from 0 to 1. This is then typically used with a threshold to assign the value to be a 0 or a 1.

- **SVM (Support Vector Machine)**: This is a classification algorithm that tries to find a vector that best separates classes. It is easy to see what this looks like in a two-dimensional space (*Figure 2.11*), but the algorithm is known to work well in high dimension spaces. Another benefit of SVM is its ability to handle non-linearity by using non-linear kernel functions, which can be used to linearize the problem:

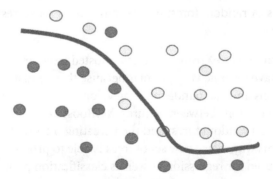

Figure 2.11 – Targets for supervised learning algorithms

- **GA2M (Generalized Additive Model)**: This is one of those rare algorithms that offers understandability, while also offering high accuracy even in a non-linear problem. The number "2" in the name represents its ability to model interactions between features. GAM model output is a summation of outputs of the effects of individual features that have been binned. Since GAM allows these effects to be non-linear, it can capture the non-linear nature of the problem. The results of the model can be represented as a simple table that shows you the contribution of each feature to the overall answer. This type of table representation is easily understandable by most people. For industries or use cases where understandability and explainability are very important, this is perhaps one of the best options you can choose.

- **K-Nearest Neighbors**: This is a very straightforward algorithm that finds the k closest data points (based on a specific way of computing distance). Now it finds the classification answers for these k points. It then determines the answer with the most votes and then assigns that as the answer. The default distance metric used is **Euclidian** distance, but DataRobot chooses the appropriate metric based on the dataset. A user can also specify a specific distance metric to be used.

- **TensorFlow**. TensorFlow is a deep learning model that is based on deep neural networks. A deep neural network is one that has hidden deep layers made up of ensembles of artificial neurons. The neurons carry highly non-linear activation functions that allow them to fit highly non-linear problems. These models are very good at producing high accuracy without the need for feature engineering, but they do require a lot more training data as compared to other algorithms. These models are generally considered very opaque and are prone to overfitting and are therefore not suitable for some applications. They are especially successful for applications where the features and feature engineering are hard to extract, for example, image processing. These models can be used for regression as well as classification problems.

- **Keras Neural Network**: Keras is a high-level deep learning library built on top of TensorFlow that allows many types of deep learning models to be incorporated into DataRobot. Being a higher-level library, it makes building a TensorFlow model a lot easier. Everything described in the preceding section applies to Keras. The particular implementation in DataRobot is well suited for sparse datasets and is particularly useful for text processing and classification problems.

Unsupervised learning

Unsupervised learning problems are those where you are not provided with an answer or a label. Examples of such problems are clustering or anomaly detection. DataRobot does not offer much for these problems, but it does have some capability for anomaly or outlier detection. These are problems where you have data points that are unusual in a way that happens very rarely. Examples include fraud detection, cybersecurity breach detection, failure detection, and data outlier detection. DataRobot allows you to set up a project without a target and it will then attempt to identify anomalous data points. For any clustering problems, you should try to use Python or R to create clustering models.

Reinforcement learning

Reinforcement learning problems are where you want to learn a series of decisions to be taken by an agent such that you achieve a certain goal. This goal is associated with a reward that is given to the agent for achieving the goal either completely or partially. There is no dataset available for this training, so the agent must try multiple times (with different strategies) and learn something on each attempt. Over many attempts, the agent will learn the strategy or rules that produce the best reward. As you can now guess, these algorithms work best when you do not have data, but you can experiment repeatedly in the real world (or a synthetic world). As we discussed before, DataRobot is not a suitable tool for such problems.

Ensemble/blended models

Ensembling is a technique for creating a model that aggregates or blends predictions of other models. Different algorithms are sometimes able to exploit different aspects of the problem or dataset better. This means that many times, you can increase prediction accuracy by combining several good models. This, of course, comes with increasing complexity and cost. DataRobot offers many blending approaches and, in most circumstances, builds the blended model automatically for your project. You can then evaluate whether the increase in accuracy is enough to justify the additional complexity.

Blueprints

In DataRobot, every model is associated with a blueprint. A blueprint is a step-by-step recipe used by DataRobot to train a specific model. See *Figure 2.12* for an example:

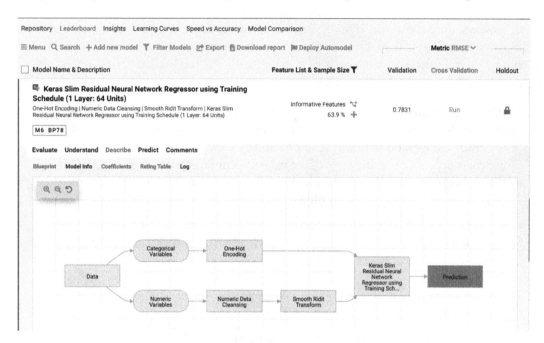

Figure 2.12 – Model blueprint

The blueprint shows all the steps taken by DataRobot to build that specific model, including any data preparation and feature engineering done by DataRobot. Clicking on any specific box will show more details on the actions taken, parameters used, and documentation of the particular algorithm used. This also serves as great documentation for your modeling project that is automatically created for you.

Now, let's look at how to determine how well an algorithm did. For this, we will require some performance metrics.

Performance metrics

DataRobot offers a wide range of performance metrics for the models. You have to specify the metric you want to use to optimize the models for your project. Typically, the best metric to use is the one recommended by DataRobot. DataRobot does compute the other metrics as well once the model is built, so you can review the results of your model across multiple metrics. Please keep in mind that no metric is perfect for every situation, and you should be careful in selecting the metric for evaluating your results. Listed here are some details regarding commonly used metrics:

- **RMSE (Root Mean Squared Error)**: RMSE is a metric that first computes the square of errors (the difference between actual and predicted). These are then averaged over the entire dataset and then we compute a square root of that average. Given that this metric is dependent on the scale of the values, its interpretation is dependent on the problem. You cannot compare RMSE for two different datasets. This metric is frequently used for regression problems when the data is not highly skewed.

- **MAPE (Mean Absolute Percentage Error)**: MAPE is somewhat similar to RMSE in the sense that it first computes the absolute value of the percentage error. Then, these values are averaged over the dataset. Given that this metric is scaled in terms of percentage, it is easier to compare MAPE for different datasets. However, you have to be mindful of the fact that the percentage error for very small values (or zero values) tends to look very big.

- **SMAPE (Symmetric MAPE)**: SMAPE is similar to MAPE, but addresses some of the shortcomings discussed above. SMAPE bounds the upper percentage value so that errors from small values do not overpower the metric. This makes SMAPE a good metric that you can easily compare across different problems.

- **Accuracy**: Accuracy is one of the metrics used for classification problems. It can be represented as follows:

Accuracy = number of correct predictions/number of total predictions

It is essentially the ratio of the number of correct predictions and all predictions. For unbalanced problems, this metric can be misleading, hence it is never used by itself to determine how well a model did. It is typically used in combination with other metrics.

- **Balanced Accuracy**: Balanced accuracy overcomes the issues with accuracy by normalizing the accuracy across the two classes being predicted. Let's say that the two classes are A and B:

(a) *Accuracy rate for A = number of correct A predictions/total number of As*

(b) *Accuracy rate for B = number of correct B predictions/total number of Bs*

(c) *Balanced accuracy = accuracy rate for A + accuracy rate for B/2*

Balanced accuracy is essentially the average of the accuracy rate for A and the accuracy rate for B.

- **AUC (Area Under the ROC Curve)**: AUC is the area under the **ROC (Received Operator Characteristic)** curve. This metric is frequently used for classification problems as this also overcomes the deficiencies associated with the accuracy metric. The ROC curve represents the relationship between the true positive rate and the false positive rate. The AUC goes from 0 to 1 and it shows how well the model discriminates between the two classes. A value of 0.5 represents a random model, so you would want the AUC for your model to be greater than 0.5.

- **Gamma Deviance**: Gamma deviance is used for regression problems when the target values are gamma-distributed. For such targets, gamma deviance measures twice the average deviance (using the log-likelihood function) of the predictions from the actuals. A model that fits perfectly will have a deviance of zero.

- **Poisson Deviance**: Poisson deviance is used for regression problems when the aim is to count data that is skewed. It works in a way that is very similar to gamma deviance.

- **LogLoss**: LogLoss (also known as cross-entropy loss) is a measure of the inaccuracy of predicted probabilities for a classification problem. A value of 0 indicates a perfect model, and as the model becomes worse, the logloss value increases.

- **Rsquared**: Rsquared is a metric used for regression problems that tells how well the fitted line represents the dataset. Its value ranges between 0 and 1. 0 indicates a poor model that explains none of the variation, while a value of 1 indicates a perfect model that explains 100% of the variation. It is one of the most commonly used metrics, but it can suffer from the problem that you can increase it by adding more variables without necessarily improving the model. It is also not suitable for non-linear problems.

Now that we have discussed some of the commonly used metrics, let's look at how to look at other results to assess the quality of your model, and the effects of different features on your model.

Understanding the results

In this section, we will discuss various visualizations of metrics and other information to understand the results of the modeling exercise. These are important visualizations that need to be inspected carefully in addition to looking at the model metrics discussed in the previous section. These visualizations are generated automatically by DataRobot for any model that it trains.

Lift chart

The lift chart shows how effective the model is at predicting the target values. As the number of data points is typically very large to show in one graphic, the lift chart sorts the output and aggregates the data into multiple bins. It then compares the averages of predictions and actuals in each bin (*Figure 2.13*):

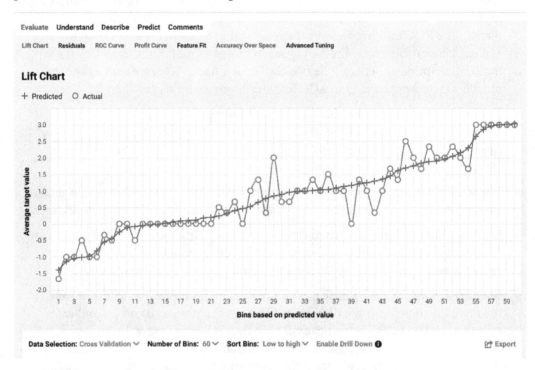

Figure 2.13 – Lift chart

The preceding lift chart shows how the predictions have been sorted from low to high and then binned (60 bins in this case). You can now see the average prediction and average actual value in each bin. This gives you a sense of how well the model is doing across the entire spectrum. You can see whether there are ranges where the model is doing worse. If the model is not doing well in a range that is important to your business, you can then investigate further to see how you can improve the model in that range. You can also inspect different models to see whether there is a model that does better in the region that is more important. Lift charts are more meaningful for regression problems.

Confusion matrix (binary and multiclass)

For classification problems, one of the best ways to assess model results is by looking at the confusion matrix and its associated metrics (*Figure 2.14*). This tab is available for multiclass problems:

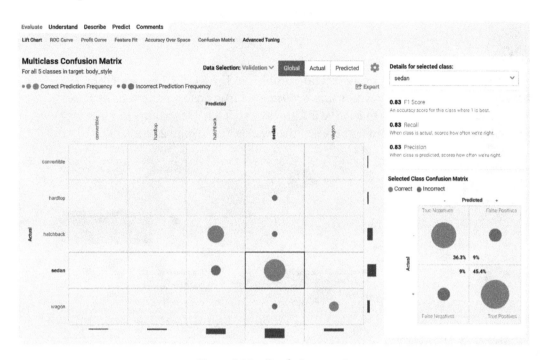

Figure 2.14 – Confusion matrix

The confusion matrix maps predicted versus actual counts (frequency) for each class. Let's look at the sedan column. The big green circle indicates how many times we correctly classified a sedan as a sedan. In that column, you will also see red dots where the model predicted it to be a sedan, but it is a different type. You can see these for all classes. The relative scales should give you an idea of how well your model did and where it is having difficulty.

If you select a specific class, you can look at the class-specific confusion matrix on the right. You can see two columns (+ for predicting a sedan, - for predicting something that isn't a sedan). Similarly, you see two rows (+ where it is a sedan, and - for when it is not a sedan). You also see some critical definitions and metrics:

- **True Positives** (**TP**) = Where it is a sedan and is predicted as a sedan
- **False Positives** (**FP**) = Where it is not a sedan but is predicted as a sedan
- **True Negatives** (**TN**) = Where it is not a sedan and is predicted as not being a sedan
- **False Negatives** (**FN**) = Where it is a sedan but is predicted as not being a sedan

Using these, we can now compute some specific metrics for this class:

- *Precision = correct fraction of predictions = TP/All Positive Predictions = TP/(TP+FP)*

- *Recall = correct fraction of actuals = TP/All Positive Actuals = TP/(TP+FN)*

- *F1 Score = harmonic mean of precision and recall. So, 1/F1 = 1/Precision + 1/Recall*

ROC

This tab is available for binary classification problems. The **ROC** (**Receiver Operator Characteristic**) curve is the relationship between the true positive rate and the false positive rate. The area under this curve is known as AUC. It goes from 0 to 1 and it shows how well the model discriminates between the two classes (*Figure 2.15*):

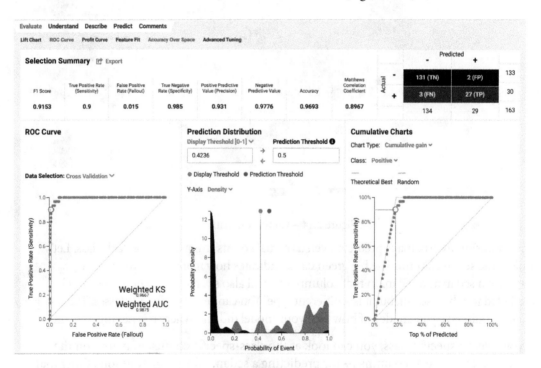

Figure 2.15 – ROC curve and confusion matrix

You can also see the confusion matrix (described earlier) and the associated metrics for the two classes. You can move the thresholds and assess the resulting trade-offs and cumulative gains. Since most problems are not symmetric in the sense that true positives have different business values compared to true negatives, you should select the threshold that makes sense for your business problem.

Accuracy over time

This tab is available for time series problems (*Figure 2.16*) and compares the actual versus predicted values over time for a series:

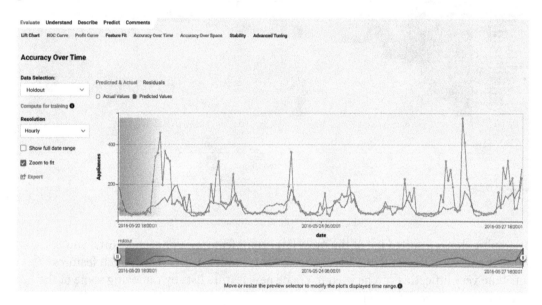

Figure 2.16 – Model accuracy over time

You can view these values for the backtests or the holdout datasets. The diagram will clearly show where the model is not performing well and what you might want to focus on to improve your model.

Feature impacts

Besides model performance, one of the first things you want to understand is how impactful the features are in terms of your model's performance. The **Feature Impacts** tab (*Figure 2.17*) is perhaps the most critical for understanding your model:

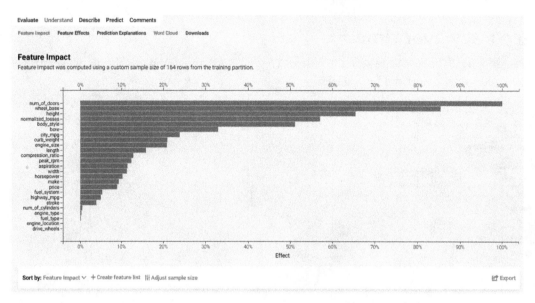

Figure 2.17 – Feature impacts

The graphic shows a sorted list of the most important features. For each feature, you can see the relative impact that a feature has on this model. You can see which features contribute very little; this can be used to create new feature lists by removing some of the features that have very little impact.

Feature Fit

The **Feature Fit** tab (*Figure 2.18*) shows an alternative view of the contribution of a feature. The graphic shows the features ranked by their importance:

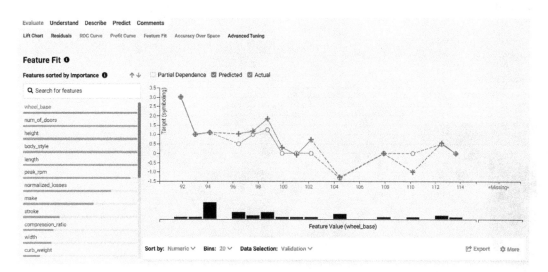

Figure 2.18 – Feature Fit

For the selected feature, it shows how the predictions compare to actuals for the range of values of a feature. Reviewing these graphs for the key features can provide a lot of insight about how a feature impacts the results and range of values that perform better and ranges where it performs the worst. This could sometimes highlight the regions where you might need to collect more data to improve your model.

Feature Effects

Feature Effects show information that is very similar to **Feature Fit** (*Figure 2.19*). In this graphic, the features are sorted by **Feature Impacts**. Also, **Feature Effects** are focused on partial dependence:

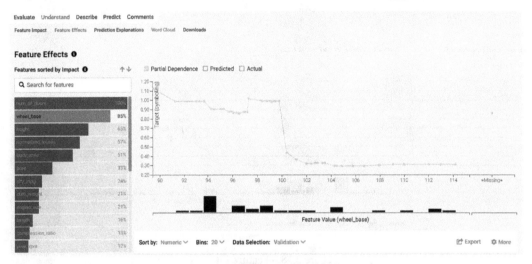

Figure 2.19 – Feature Effects and Partial Dependence

Partial dependence plots are one of the most important plots that you want to study carefully. These plots tell you how a change in the value of a feature impacts the change in the average value of the target over a range of values for the other features. This insight is critical to understanding the business problem, understanding what the model is doing, and, more importantly, what aspects of the model are actionable and what range of values will produce the maximum impact.

Prediction Explanations

Prediction Explanations describe the reasons for a specific prediction in terms of feature values for the specific instance or row that is being scored (*Figure 2.20*). Note that this is different from **Feature Impacts**, which tell you the importance of a feature at a global level:

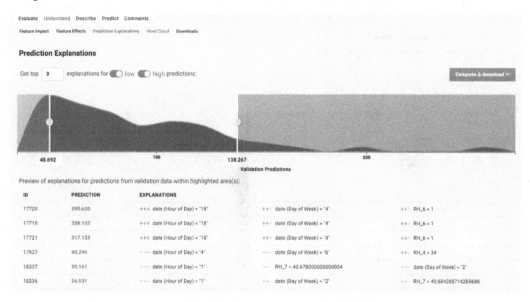

Figure 2.20 – Prediction Explanations

Prediction Explanations can be generated for an entire dataset or a subset of data, as shown in the preceding screenshot. For example, it will provide the top three reasons why the model predicted a specific value. These explanations are sometimes required for regulatory reasons in certain use cases, but it is a good idea to produce these explanations as they do help in understanding why a model predicts a certain way and can be very useful in validating or catching errors in a model. DataRobot uses two algorithms for computing the explanations: **XEMP (exemplar-based explanations)** or **Shapley values**. XEMP is supported for a broader range of models and is selected by default. Shapley values are described in the next section.

Shapley values

Shapley values (**SHAP**) are an alternative mechanism for producing prediction explanations (*Figure 2.21*). If you want to use SHAP for explanations, you have to specify this in the advanced options during the project setup before you press the **Start** button. Once DataRobot starts building the models, you cannot switch to SHAP. SHAP values are only available for linear or tree-based models and are not available for ensemble models:

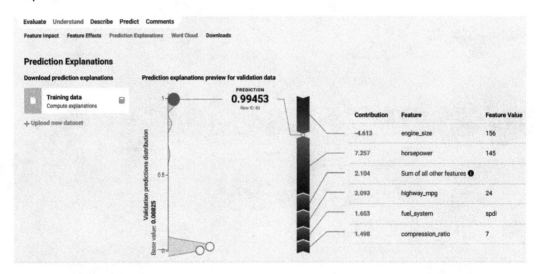

Figure 2.21 – SHAP-based explanations

SHAP values are based on cooperative game theory, which tries to assign values to contributions of a team member in a collaborative project. In the context of machine learning, it tries to assign the value contribution of a specific feature when there is a team of features collaborating to make a prediction. SHAP values are additive and you can easily see how much of the final answer is due to a specific feature value.

Summary

In this chapter, we covered some of the basic machine learning concepts that will come in handy as we go through the remaining chapters, and they will also be useful in your data science journey. Please note that we have only covered concepts at a high level, and depending on your job role, you might want to explore some areas in more detail. We have also related this material to how DataRobot performs certain functions and where you need to pay closer attention.

Hopefully, this has given you some insights into what DataRobot will be displaying and where to focus your attention in different stages of your project. Since DataRobot automates a good chunk of model building and prediction tasks, it might be tempting to ignore many of the outputs that DataRobot is automatically producing for you. Please resist that temptation. DataRobot software is taking considerable pains and resources to produce those outputs for a very good reason. It is also doing much of the grunt work for you, so please take advantage of those capabilities. Specifically, we have covered the following: What are the things to watch out for during data preparation? What data visualizations are important for gaining an understanding of your dataset? What are the key machine learning algorithms, and when do you use them? How do you measure the goodness of your model results? How do you assess model performance and understand what the model is telling you about your problem?

Now that we know the basics, we will start our data science journey in the next chapter by learning how to understand the business problem and how to turn it into a specification that can be solved by using machine learning.

3
Understanding and Defining Business Problems

This chapter covers topics that are the most critical for success and yet are not discussed in detail in data science programs or books. Although the topic of understanding and defining business problems is mentioned very briefly as something that should be done, it is very rare that the discussion will go into how to actually do it properly. In this chapter, we will go into specific tools and methods that can be used to gain an understanding of the system under consideration and determine the problem that needs to be solved.

This section is independent of DataRobot, as DataRobot cannot help you with this part of the process. This is something that a data analyst, a business analyst, or a data scientist has to do. Correctly defining a business problem is hard to do—it is not automatable, and it is also not done properly most of the time. If you gain this skill, you will become invaluable. This is a key area where there will always be a need for experienced data scientists (or whatever they are called in the future).

By the end of this chapter, you will have learned about some of the core concepts and methods you need to know in order to ensure that you are solving the right problems. The rest of the book will explain how to solve those problems in the right way.

In the chapter, we're going to cover the following main topics:

- Understanding the system context
- Understanding the *why* and the *how*
- Getting to the root of the business problem
- Defining the **machine learning** (**ML**) problem
- Determining predictions, actions, and consequences for Responsible **artificial intelligence** (**AI**)
- Operationalizing and generating value

Understanding the system context

All problems arise within the context of a system. A system could be a single cell of an organism, a global population, or the entire economy. In the same way, all solutions need to fit into a system. A technological solution (for example, an AI solution) will typically require changes to processes, people, skills, other IT systems, or even the business model for it to be effective. For an organization, the system could be its entire supply chain, competitors, and customers. Given that a system's definition can be very broad, it is generally advisable that you imagine a system to be broader than the problems you are considering. You want all the components or agents that your problem touches to be part of the system context. Defining the system boundary is part art and part science, and it is an iterative process. Given that you will be looking at the system from a broader perspective, this also means that the same system context will be valid for multiple ML projects or use cases. The understanding you gain here will pay dividends across many projects.

Data scientists or analysts who might have worked in an organization or industry would have intuitively learned many of the systemic aspects of the problem. They might feel that they do not need to look into this further as they already understand key issues. While that might be true, it is also true that people develop blind spots and start to overlook key missing pieces or carry implicit assumptions that are mostly correct but sometimes wrong. Using structured methods to capture systemic understanding helps overcome these problems and also ensures that everyone is working from a common understanding. These issues are typically ones leading to problems or delays in projects downstream. Let's look at how we build this understanding by creating a context diagram.

A context diagram is a high-level view of your system, showing key players and their interactions, as illustrated here:

Figure 3.1 – Context diagram

The specific diagramming convention is not that important; what is more important is that you understand and document the components and understand how they interact. There are many diagramming conventions out there, so feel free to use the one you like. Make sure to capture three to five important instances of each topic in the diagram. The arrows need not be one-way.

As you look at this simple diagram, you will agree that we should know all of these things. As you try to build this diagram, you might be shocked to learn that finding and capturing this information is not that straightforward. Most people in an organization will have some notion of these things but might not be able to precisely specify the most important customers or **key performance indicators** (**KPIs**), and so on. In most organizations, it might take some time and discussion to put this together. Most of the components in the diagram are easy to understand but some are a bit confusing, so it's worth discussing them a little bit, as follows:

- **Key objectives**: Key objectives are measurable metrics that let you determine whether you are achieving your goals in a timely fashion. These typically take the form of financial performance, market share, customer satisfaction, reputation, quality levels, and compliance. It is important to have precise and measurable definitions and alignment with the goals of the organization. These represent true value to an organization, and it is important to understand how your projects and models impact these.

- **External drivers/risks**: These are external factors or drivers that impact the key objectives but are not under your control. Notice that we are not discussing specific events, but changes in value of factors that might be considered events—for example, the factor might be **gross domestic product (GDP)** change. It is not in our control, and a value of -20% might indicate a financial crash. So, our driver in this case is GDP change as opposed to financial crash.

- **Key decision levers**: These are also drivers that impact key objectives, but they are in our control. For example, the number of employees is a factor that is in our control (as we can decide how many to hire) and it will have an impact on outcomes. Other examples could be how much to invest in new technology or in marketing, and so on. These could be strategic decisions or choices, such as creating a new distribution channel, bundling products, and so on. Regardless of type, the important thing to remember is to make sure that the idea is captured in a precise way—for example, if a new distribution channel is a driver, you should know what the five actual choices are.

> **Note of caution**
>
> Please do not get trapped in philosophical debates. Quickly create the first iteration of your diagram and refine it in the future, as needed. It is OK to move forward with the first draft, as your analysis might inform and change the current thinking.

As you may have guessed by now, the reason for highlighting these three items is that historic data about them will be critical for any data science project. You will also agree that data about these factors is critical for operational as well as strategic decisions, yet you might find that this data might not be easily available or might have quality issues. In addition, pay special attention to key knowledge stores. These will be the databases, data warehouses, data lakes, or filesystems that contain data for your organization. We will revisit these items again in the following sections.

Now that we understand the context, we want to understand how our system operates and why it behaves a certain way. Both of these aspects are critical to understanding the system. In the next section, we will describe how to create that understanding.

Understanding the why and the how

The key to understanding a system's function and its behavior lies in the following aspects:

- **Process**: How do objects and information flow through the system's processes?

- **Interaction**: How do different entities or components of the system interact with each other?

- **State**: How does the state of an entity evolve over time?

- **Causal**: What are the causal relationships?

Each of these aspects is represented via diagrams. There are many diagramming conventions for process modeling, causal modeling, and interaction diagrams. These conventions are used differently in different domains. You can follow any of the conventions that you like or are already familiar with. In this book, we will follow certain conventions that are amenable to computational modeling. What that means is that these diagrams can be combined with data and turned into models that can be used for analysis. This will become important in the later stages of our project. In-depth details of how to create these diagrams and turn them into computational models are beyond the scope of this book, but if you are interested you can seek out other sources to learn about these techniques. Even if you do not create computational models, these diagrams will provide useful insights that you will be able to use in your ML projects. It is possible that you have people in your organization that build—or have built—such diagrams. You should seek those people out and elicit their help in building these diagrams. Let's look at each of these aspects in more detail.

Process diagrams

In an organization's operation, there are several functions that have well-defined processes. Objects, people, or information flow through these processes, as illustrated in the following diagram:

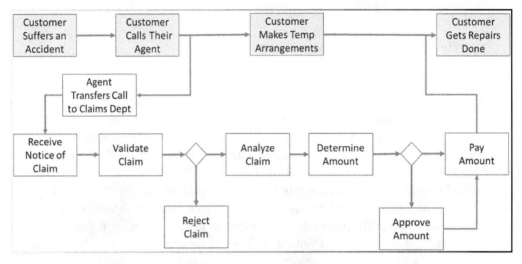

Figure 3.2 – Process diagram

Based on the system context, you already know the most important processes in an organization. You can start building diagrams for these key processes—or at the very least, the one that is relevant to your project. It is important to build end-to-end process flows—for example, the entire customer journey, or the entire development process for a product. It is also important to understand the process from a customer's viewpoint (gray boxes) and not just internal processes (white boxes). Make sure to capture failure points or rework paths, or where a process might end abruptly instead of normal completion. As entities flow through these processes, decisions or computations are made that could be candidate ML problems. Regardless of the project you are starting from, it is a good idea to identify other potential opportunities along the way. It may turn out that looking at the process differently or building a different ML model could provide larger benefit or might preempt a need for the current project. Whether or not that turns out to be the case, it is important to capture this information. By the way, did you notice in the preceding process diagram that the customer is not receiving a reject message? As you can imagine, this is an important part of the customer experience that is being left out. I am sure that error will be caught at some point, but making the process explicit increases the odds of catching it sooner and taking it into account as you are building your models.

Besides building the diagram, it is important to capture data about the process. You will frequently run into situations where someone has already built a process diagram but did not capture any data. If you are reading this book, then I do not need to tell you how important collecting that data is for accurately understanding the process. Typical information to be captured is counts and types of objects flowing through each step, time taken at each step, labor hours and resources required at each step, probability of taking a specific path, quality metrics, and so on. If such information is not being captured for key processes, then it is important to start collecting this information as soon as possible. This information could be critical for building a business case for your project, serving as useful features in the model, and helping identify problems that might be otherwise hidden.

It is important to note again that you do not want to get stuck in terminology debates, and instead quickly create a diagram that is sufficient to help you understand what happens in the process, as opposed to a very detailed view with every little nuance in it. It is OK to revisit this if a need arises to get into more details on some specific aspect of the process.

Actual processes are typically a little more complex than what we show in *Figure 3.2* but not by much, and it is not uncommon to uncover things that are not known to many people outside the specific department where these tasks are done. In my experience, it is also not uncommon to find that no single person understands the entire process. The exercise can be valuable in itself by highlighting key problems but it is especially valuable to the data science team building models to automate some part of this process, yet I am surprised to see how many times data science teams build models without understanding the process.

Interaction diagrams

There are many interactions happening in a system that do not follow a fixed process. These interactions can happen in different orders and need to be kept flexible, and are best understood via interaction diagrams. Those of you with software development backgrounds are likely familiar with such diagrams that show interactions among software components or objects. In our case, we are interested in understanding interactions between key players in a system, as illustrated in the following diagram:

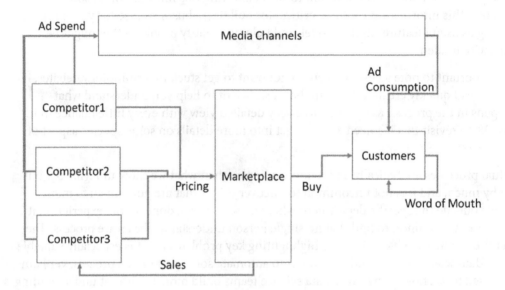

Figure 3.3 – Interaction diagram

The diagram shows a marketplace where several competitors are selling their products. Customers come to this marketplace to purchase the products. The competitors spend money to advertise their products on various media channels, and set their prices. The consumers are influenced by the advertising, pricing, and word of mouth from other consumers. At any given time, many of these interactions are taking place, creating a complex and dynamic environment. If you are building a pricing model you have to take all of this into account, or your model will show great statistical fit to data but will prove ineffective during operation.

Note that key players can be people, organizations, bots, marketplaces, fraudsters, and so on. The idea behind building this and other diagrams is to codify and make explicit what you know. This enables everyone to share a common view of the system and question assumptions or point out missing information. It is also important to make sure that you treat these diagrams as hypotheses that need to be tested with data. You have to continually ask whether the data supports what we are saying in these diagrams. If not, then maybe your assumptions need to be refined, or perhaps you have missing data or other data-quality problems. Perhaps the data collection is biased. I am sure you have heard stories about how biased data was used to make predictions that turned out to be totally wrong. Building a diagram is not a guarantee that you will catch biased data, but it does improve your odds of catching it.

State diagrams

A state diagram captures the evolution of state of some important entity or actor in a system. Typical candidates are customers and products. As with other diagrams, you build these diagrams for only the important or critical actors in a system.

An example state diagram can be seen here:

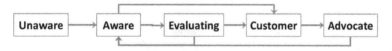

Figure 3.4 – State diagram

State diagrams are very similar in concept to **Markov chains** (this concept represents the probability of transitioning a system from one state to another state in such a way that the probability is fixed and not dependent on any previous history), except in a state diagram you do not have to assume that history does not matter. A state diagram is built for a specific agent. *Figure 3.4* shows a state diagram of a person progressing through various states over time. The arrows represent transitions from one state to another, and the person stays in a given state until they transition to the next state. The transition is typically assumed to be instantaneous. You can also think of the transitions in terms of transition probabilities (in which case, it starts resembling a Markov chain). The diagrams can be hierarchical in the sense that a state can be decomposed into sub-states, and those sub-states can be interconnected via transitions.

In addition to building a diagram, you want to understand what causes a transition to take place. Is it deterministic or is it random? You also want to collect data about how often and when these transitions take place as this data is very useful for further analysis, as well as for building ML models. Transitions of one actor might cause a transition in another actor's state, thus state diagrams are also connected to interaction diagrams. Each of the transitions is a potential candidate for an ML model, where you can use data to predict when a transition (and which one) might fire. As you can imagine, building these diagrams will lead to the identification of opportunities that might otherwise be missed.

Using these diagrams, you now have an understanding of how a system functions. We are now ready to look at what determines a system's behavior.

Causal diagrams

Causal diagrams intend to capture our understanding of cause-and-effect relationships present in a system. This understanding may or may not be correct. In fact, you might never be able to prove causation. Philosophical debates aside, you can greatly improve your understanding by using the methods outlined in this section, combined with data.

An example causal diagram can be seen here:

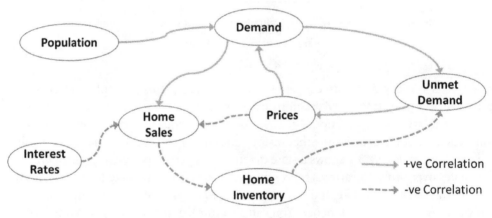

Figure 3.5 – Causal diagram

The preceding diagram shows the relationships in a housing market (this is just an example—it doesn't show all the causal influences). It says that as interest rates go up, home sales decline. Home sales are also influenced by demand and prices. As prices go up, the rising prices can increase demand as more people want to buy homes to make money, but the price itself is a deterrent. You can see that there are opposing effects and feedback loops present in this simple diagram. No wonder the dynamics confound us, and this frequently leads to the system going haywire. Everyone thinks they understand how the housing market works, but real understanding is difficult to achieve in the presence of complex dynamics. Similar dynamics are at play in many business situations. It is easy to build an ML model to predict home prices, but it is much harder to understand the overall dynamics. This lack of understanding can lead to a situation where you are using excellent ML models to make bad decisions, hence building such models is critical for understanding the true nature of the problem you are trying to solve. Such diagrams are also useful for understanding and treating confounding variables and counterfactual analysis [Pearl].

There is one more representation of causal models that comes from the discipline of system dynamics. This representation combines some ideas from the other diagrams with causal diagrams to create a view that can be very insightful and can be easily turned into a dynamic simulation model. System dynamics is a vast discipline in itself, and there are many good books and papers on the topic [Sterman]. Here, I will only introduce the notion and what it looks like, and how it can be useful. Here is an example system dynamics (also known as stock and flow) diagram:

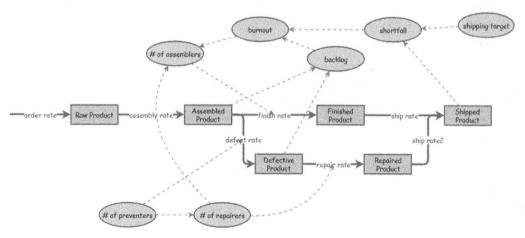

Figure 3.6 – System dynamics (stock and flow) diagram

This diagram captures ideas from a state diagram, a process diagram, and a causal diagram into a composite view that can be very instructive. Imagine the journey of a product as it changes state, going from a *raw product* to a *shipped product*. Each box in this case represents the quantity of a product in a specific state—for example, the **Assembled Product** box represents all products that have been assembled and are now waiting to become *finished products*. This happens at a rate called the *finish rate*. The finish rate is dependent on how many assemblers are available to perform the work. You will also notice that some assembled products turn out to be defective. These products flow into the **Defective Product** box at a rate called the *defect rate*. These then have to be repaired by repairers. Because of defects, the shipping target is not met and there is a shortfall. This shortfall increases pressure on the employees, which increases burnout. The burnout in turn reduces the number of assemblers (they quit or get sick). Since the number of assemblers is reduced, this slows down the finish rate. This in turn increases the pressure and more people are shifted to become repairers, as repairs can be done faster. This leaves no one working as preventers and causes the vicious cycle to continue, with the process becoming more and more backlogged.

This dynamic plays out in many organizations and they wonder why they are always under pressure. Once the diagram is laid out, you can see the problem is that they are fixing symptoms as opposed to the root problem of defects. In this simple diagram it is easy to see, but in more complex situations you can run simulations of these diagrams to find the problem points. The diagram also helps to clarify the relationship between the processes, decisions, and business objectives. These diagrams can be simulated to understand the business impacts of decisions, as well as the impacts of deploying ML models. This is a great way to show the value of your efforts in a way that most people can easily understand.

Now, let's come back to ML. If this analysis is not done, then it is most likely that defect repair will get flagged as a problem, and it is likely you will be building a model to predict how many defects will be created or predict how many items will be shipped. You will build a great model, but that will not solve the problem. The key problem is to find which factors are causing the defects and how the defects can be reduced. This will require the manufacturing team to work closely with the data science team to find a solution. Again, the key point is that unless someone does this analysis, the data science team is likely to be solving the wrong problem. You might think that this doesn't happen often, but I contend that this happens a lot more than you might think because the true problem often stays hidden for a long time.

In general, it is best to treat each causal relationship you have drawn in these diagrams as a hypothesis. The diagram then represents a collection of hypotheses. There are statistical and simulation methods that can be used to validate these hypotheses. For this, you will need to start collecting data or start discovering where that data is stored. Now that we have learned about these diagrams, let's look at how we get to the root of the problem.

Getting to the root of the business problem

Some problems are easy to solve, while others prove to be much harder. One of the reasons this happens is that when a problem's symptoms appear somewhere else and after some delay, then it is very difficult to know where the problem really is. By definition, the symptoms are clearly visible—they are explicit and you can easily collect data about them. The underlying problem, on the other hand, is happening in some other department or building and is not visible because it is not causing immediate pain. Most likely, no data is being collected about the root problem, or it might be too hard to collect that data. Given the nature of ML, it is almost a given that all the data you are getting is about symptoms. If you are lucky, you might get some data about the root problem as well (although you will not know it).

One of the ways to get started is by using an old method called *five whys*, which basically involves asking the question *Why?* five times to get to the root cause. Many times, there might be multiple causes at each level. So, in practical terms, many people use another diagram that is called a *fishbone diagram* to capture this information, as illustrated here:

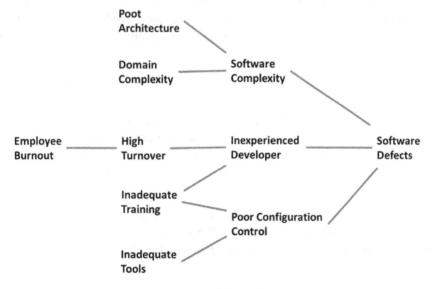

Figure 3.7 – Fishbone diagram

In the preceding fishbone example diagram, we are trying to understand why we have a software defect problem. As you ask *why* questions and capture the causes, you begin to add potential causes. You continue this process until you feel that you have captured the essence of the problem. As you can tell, this is partly a causal diagram and can be used as a starting point for building a causal model for your system while maintaining focus on a specific problem. Data can then be collected to confirm or reject the hypotheses. As you can see, this fishbone diagram will inform the system dynamics diagram we saw in *Figure 3.6*. The key point of the exercise is to understand the root cause and work on fixing this. I hope you are beginning to see that accurately predicting something is not always the same as fixing the real problem, and you have to be a bit careful about setting up your ML problem for your project to be successful. Also, notice that there might be multiple causes leading up to a problem and you might have to address multiple issues in order to see significant mitigation of the problem. On the other hand, addressing only one aspect (say, employee burnout) without understanding the full picture could lead you to conclude that burnout is not a problem. It is not uncommon in organizations to have reached wrong conclusions that then get incorporated into the myths of the organization—"*we already tried that, and it does not work.*"

Now that we have identified the business problem that needs to be tackled, let's look at how we turn it into an ML problem.

Defining the ML problem

Before we get into the details of ML, it is important to note that not all problems are appropriate to be solved with ML. For a problem to be a good candidate, it should have the following characteristics (we will focus only on **supervised learning** (**SL**) problems for now):

- There is a clear target or label value that would be useful if an algorithm can predict it. In the absence of an algorithm this value remains unknown, requires a person's judgment, or requires substantial effort for it to be determined. Sometimes, the target will not be the actual variable of interest but a critical component of that calculation. This part is not always obvious, but the problem analysis you did in previous sections of this chapter will certainly help in clarifying which variable makes the best target.

- You have access to a large enough historical dataset that contains the values of the target or label you wish to predict. You will need to create a list of data sources that contain relevant data and start understanding the data that they contain.

- Determine which type of SL problem is best suited to your problem (regression, classification, or time series). Sometimes, you can cast one type of problem into another type.

- There is typically a trade-off between accuracy, explainability, and understandability. It is important for you to consider what is more important for your business problem. In many situations, we are willing to sacrifice accuracy to improve understandability. This, in turn, determines which algorithms and which explanatory method you should select.

You will have to review your datasets and the business problem definition to see whether you can craft a specification such that the conditions listed previously are satisfied. In doing so, there are several transformations that might have to be performed, such as these:

- Transform the target such that it is more valuable as an output or better suited to solve the business problem. Take the following examples:

  (a) If numeric target values are over a very large range of values or are distributed with a high skew, then you can try to log the target as a new target.

  (b) If the actual value is less important than a range, then you can create bins and use the binned values as targets.

  (c) Sometimes, a change in value or the rate of change makes for a better target.

- Create interaction features based on the causal diagrams. Take the following examples:

  (a) Intermediate variables in the causal model that are not directly observed but can be computed by combining various other features.

  (b) If your target is in log form, then it might make sense to create logs of various features.

  (c) Similarly, you might want to bin certain features if the bins have a special significance in the business problem.

  (d) Explore whether creating rate-of-change features would be important for your model.

- Identify missing data that showed up in the causal diagram but is not present in your dataset. Depending on the importance and how easy it is to get this, you might want to collect it before building the model. Another choice is to get the data collection process started in parallel with building the model with the data you already have. In the latter case, you can always revisit the model in the next iteration when you have collected the data.

In addition, you also have to think about how you will define and assess errors. Which metric will be best for the problem? We covered metrics in *Chapter 2*, *Machine Learning Basics*, and DataRobot automatically picks an appropriate metric for the problem. I have found the selections to be very good most of the time, so it is a good idea to go with the recommendation unless you have a good reason not to. In addition to metrics, you need to think about whether you care more about errors in a specific range versus some other range—for example, maybe you want to be more accurate in the high-value range of a feature versus the low-value range. In such cases, you can consider using this numeric feature (this has to be a non-target feature) as a weight for computing the error metrics. You can find this setting under **Advanced Options** in the **Additional** tab, as illustrated in the following screenshot:

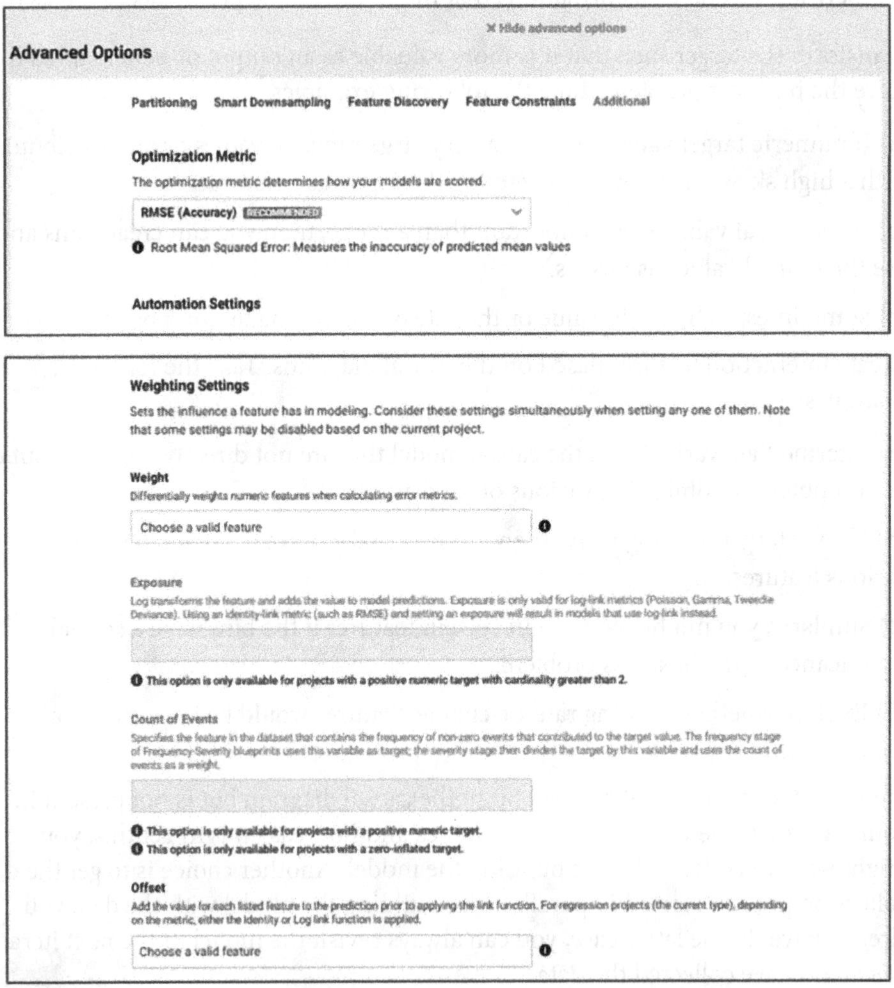

Figure 3.8 – Additional options

The preceding screenshot shows the options for selecting a feature to be used as a weighting for the predictions. Note that this has to be set at the start of a project. Once the model building starts, you cannot change this setting.

Once you have the ML problem defined, DataRobot comes into action to build the models. We will cover this in later chapters. For now, we want to discuss what happens when the models are built and are now able to generate predictions.

Determining predictions, actions, and consequences for Responsible AI

After the model is built and deployed in DataRobot, it might seem that our job is done—but not so fast. You should start analyzing what the predictions profile looks like and start discussing with users and stakeholders the details of actions to be taken. The models you have helped build are likely to introduce many changes in your system and will impact other people. It is therefore important to try to make sure that these impacts are not negative. Making sure that your models will not cause harm is called Responsible AI. This will build upon the work you did during the understanding phase through various diagrams.

Just as in previous sections we saw how a causal diagram helps you to relate features to a target, we can also see how the target affects other parts of the system. The diagram should reveal how the target impacts key objectives or outcomes; it should also reveal key feedback loops that will change the behavior of the rest of the system, as well as giving a hint of other consequences. This makes it relatively straightforward to understand and compute the **return on investment (ROI)** from your model. A common challenge expressed by data science teams is that they find it hard to express the impacts of a model. As you can now see, the work we put into understanding our system from a causal perspective also helps to determine the business impact of the model.

It is very common to see that most systems don't provide a free lunch—there are always trade-offs. Your actions might improve one objective but might hurt another. It is very important to understand these trade-offs and to ensure that your stakeholders understand them as well. It is very possible that even though the model will provide a positive ROI, it might cause performance to degrade in other areas, such that you might not want to go forward with it. Some of these consequences could be in the areas of regulatory or ethical issues. These areas are often overlooked, only to be discovered (painfully) at a later point in time. One of the key benefits of doing this analysis is to make sure you are catching any problems earlier.

Another mechanism that is often used is to simulate the system dynamics diagrams. This allows you to understand the dynamic behavior of your system and can serve as a virtual experiment. Virtual experiments or simulations let you test out different strategies in a safe environment before launching your models. Not only does this help you avoid costly mistakes—it can often also suggest improvements or strategies to further optimize the benefits. The reasons many data science projects do not succeed are that data science teams rarely do this, typically do not have the skills to do this analysis, and have historically not taken this part very seriously. Let's look here at a simple example of an ML model in a system:

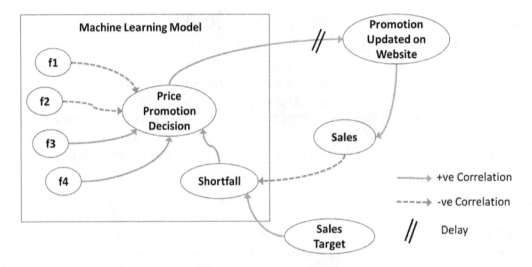

Figure 3.9 – Impact analysis

In this example, you have a price promotion model that uses some features to generate a promotion price. This model monitors the sales and updates the price accordingly. For some reason (database updates, approvals, and so on), there is a delay involved in updating these prices on the e-commerce site. This creates a lag in the sales data that is not known to the modeling team because the modeling team did not understand the entire flow and how much time it takes to update prices on the site. If you were to do an analysis of this simple diagram, you would discover that such feedback loops with delays produce oscillating behavior. This means that the pricing will always be off—sometimes higher and sometimes lower. This behavior is somewhat similar to what you see in most model results anyway, so it is entirely possible that this effect will be missed. The system will perform poorly even though the model itself was fine. I have kept this example very simple to make a point. You can imagine that if the situation were more complex, most people would not be able to see the problem on their own till it is too late.

On the flip side, you can also evaluate how much sales impact your model will make compared to the status quo. This is a great way to show ROI and the business value of your work. Since simulations based on these diagrams are a visual representation of your business, most people find it a lot easier to understand as compared to narratives or spreadsheets. This also helps in gaining acceptance of your ML models.

While DataRobot does not help with many of these aspects, it does offer mechanisms to determine whether your model is biased along with any protected features and measure the amount of bias that exists in the model. This, combined with the preceding analysis, can go a long way to ensure that your models are not biased and that the results are not being used in a way that goes against your organization's values. Now that we are happy with the model's expected results, we can start the process of deployment.

Operationalizing and generating value

Operationalizing a model in your infrastructure can be a complicated undertaking. There are some aspects of deployment that are made simple by DataRobot, but there are other parts of deployment that are outside the scope of DataRobot and can be quite challenging. Let's discuss the tasks that are part of this process, as follows:

- **Deploying a model as an application programming interface (API)**: One of the very first tasks is to deploy your model as an API so that it can serve predictions as needed. You will have to decide whether this needs to be a batch or real-time operation. DataRobot automates much of the task of setting this up, and you can have an API serving predictions in minutes.

- **Integration and testing with business systems**: Having an API is only part of the story—you will now need to integrate this API into your business systems. Sometimes, you can serve up predictions to users via standalone Excel files or web pages, but for many use cases integration is required. This can sometimes take time and effort and can slow things down. Another potential path that many organizations are beginning to use is **robotic process automation (RPA)**. DataRobot offers integrations with several RPA tools that can speed up the integration process if your use case is amenable to an RPA implementation.

- **Building an end use interface**: This is not needed if your use case calls for complete automation, but most use cases will have some level of user involvement. With integration out of the way, you will still require some way for the user to interact with the prediction and make appropriate decisions. You will need to consider how users will adjust to a new way of doing business and how to make this experience as frictionless as possible. In fact, in many use cases, the predictions are specifically set up to reduce friction in an existing process.

- **User training**: Make sure you are planning for and ready to offer training to users whose workflow is being impacted by the new models. Creating training and offering this training is a great way to increase adoption and acceptance of the model. Many times, this is thought of after the fact and can cause delays or reduce acceptance.

- **User acceptance and change management**: This is typically an ongoing process. It is generally a good idea to involve users and stakeholders from the start. If users feel that their voice is heard, this will improve the chances of acceptance. Users can also help avoid potential problems that the data science team will not catch on their own. Frequent communication about why you are doing this project, how it impacts the users, and how their work will change (hopefully for the better) are all good strategies to improve your odds of success. Building the diagrams listed previously in conjunction with users is a great way to start this dialog and is ultimately what adds value to the business. As you can see, many things have to happen before and after a model is built to realize value. It is no wonder that projects often do not succeed in adding value.

- **Model monitoring and maintenance**: Once the model is operational, you will need to set up mechanisms to monitor the prediction service and the performance. Over time, the performance tends to degrade, or you might want to improve the performance of the model. This requires the models to be updated or retrained with new data. Luckily, DataRobot makes these tasks very easy as it provides mechanisms to set up the monitoring and retraining of the models.

Summary

In this chapter, we covered some tools and methods to help you gain an understanding of your system and the business problem you are trying to solve. Some of these methods will be new or unfamiliar to even experienced data scientists, but it is important to take the time to internalize them and practice them on your projects. Some of this will feel unnecessary given the time pressures. This is one of the reasons tools such as DataRobot are beneficial, as they reduce the time you need to spend on repetitive tasks and allow you to focus on things that tools cannot do.

Hopefully, I have convinced you that the combination of data science teams focusing more on understanding the problem and using automation tools for some of the model building and tuning tasks provides the best value to an organization. A lot of the work done here will also come in handy toward the end of the project when we are getting ready to operationalize the models into the organization. Specifically, in this chapter, we have learned how to understand the broader system context, how the system operates, and why it behaves a certain way. We also saw how to get to the root problem that a business needs to solve and turn the business problem into a form that can be solved with ML. We then learned how to make sure that the solution solves the right problem and does not create unintended side effects.

Finally, we learned how to make sure that the solution is accepted by the stakeholders and gets operationalized, leading to the realization of business value.

We are now ready to start working with some example datasets and begin using DataRobot to help solve the business problem we have uncovered and the ML problem we have defined.

Further reading

- *Causality: Models, Reasoning and Inference, Second Edition*, Judea Pearl, Cambridge University Press.

- *Business Dynamics: Systems Thinking and Modeling for a Complex World*, John D. Sterman, Irwin/McGraw-Hill.

Section 2:
Full ML Life Cycle
with DataRobot:
Concept to Value

This section will cover the entire life cycle of building and deploying an ML model with DataRobot. Upon completion, you will know how to take a project from start to finish. Although the tasks are listed linearly, these tasks will happen iteratively during any real project, and at many points in this process, you will jump back to a previous step to perform some tasks all over again.

This section comprises the following chapters:

- *Chapter 4, Preparing Data for DataRobot*
- *Chapter 5, Exploratory Data Analysis with DataRobot*
- *Chapter 6, Model Building with DataRobot*
- *Chapter 7, Model Understanding and Explainability*
- *Chapter 8, Model Scoring and Deployment*

4
Preparing Data for DataRobot

This chapter covers tasks relating to preparing data for modeling. While the tasks themselves are relatively straightforward, they can take up a lot of time and can sometimes cause frustration. Just know that if you feel this way, you are not alone. This is pretty normal. This is also where you will begin to notice that things are a bit different from your experience in an academic setting. Data will almost never arrive in a form that's suitable for modeling, and it is a mistake to assume that the data you have received is in good condition and of good quality.

Most real-world problems do not come with a ready-made dataset that you can start processing and use to build models. Most likely you will need to stitch data together from multiple disparate sources. Depending on the data, **DataRobot** might perform data preparation and cleansing tasks automatically, or you might have to do some of these on your own. This chapter covers concepts and examples to show how to cleanse and prepare your data and the features that DataRobot provides to help with these tasks.

By the end of this chapter, you will know how to set up data to hand it off to DataRobot and begin modeling. In the chapter, we're going to cover the following main topics:

- Connecting to data sources
- Aggregating data for modeling
- Cleansing the dataset
- Working with different types of data
- Engineering features for modeling

Technical requirements

Some parts of this chapter require access to the DataRobot software, and some tools for data manipulation. Most of the examples deal with small datasets and therefore can be handled via Excel. The datasets that we will be using in the rest of this book are described in the following sections.

Automobile Dataset

The Automobile Dataset (source: Dua, D. and Graff, C. (2019). UCI Machine Learning Repository [http://archive.ics.uci.edu/ml]. Irvine, CA: University of California, School of Information and Computer Science) can be accessed at the UCI Machine Learning Repository (https://archive.ics.uci.edu/ml/datasets/Automobile). Each row in this dataset represents a specific automobile. The features (columns) describe its characteristics, risk rating, and associated normalized losses. Even though it is a small dataset, it has many features that are numerical as well as categorical. Features are described on the web page, and the data is provided in .csv format.

Appliances Energy Prediction Dataset

This dataset (source: Luis M. Candanedo, Veronique Feldheim, Dominique Deramaix, *Data driven prediction models of energy use of appliances in a low-energy house*, Energy and Buildings, Volume 140, 1 April 2017, Pages 81-97, ISSN 0378-7788) can be accessed at the UCI Machine Learning Repository (`https://archive.ics.uci.edu/ml/datasets/Appliances+energy+prediction#`). This dataset captures temperature and humidity data in various rooms in a house and in the outside environment, along with energy consumption by various devices over time. The data is captured every 10 minutes. This is a typical example of a time series dataset. Data is provided in `.csv` format, and the site also provides descriptions of the various features. All features in this dataset are numeric features. The dataset also includes two random variables to make the problem interesting.

> SQL
>
> For some parts of this chapter, it will be helpful to know SQL, although you do not need to know SQL to go through the example problems.

Connecting to data sources

By this point, you should have a list of data sources and an idea of what data is stored there. Depending on your use case, these sources could be real-time data streaming sources you need to tap into. Here are some typical sources of data:

- Filesystems
- Excel files
- SQL databases
- Amazon S3 buckets
- **Hadoop Distributed File System (HDFS)**
- NoSQL databases
- Data warehouses
- Data lakes
- Graph databases
- Data streams

Depending on the type of data source, you will use different mechanisms to access this data. These could be on-premises or in the cloud. Depending on the condition of the data, you can bring it directly into DataRobot, or you might have to do some preparation before you bring it into DataRobot. DataRobot has recently added capabilities in the form of **Paxata** to help with this process, but you might not have access to that add-on. Most of the processing work is done via **SQL**, **Python**, **pandas**, and **Excel**. For the purpose of this book, we will only focus on Excel.

If you are not already familiar with SQL and pandas, then it will be helpful for you to start learning about them as soon as you get an opportunity:

1. You can connect to a data source by going to the **Create New Project** menu, as shown in the following figure:

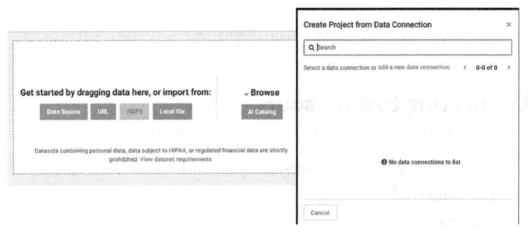

Figure 4.1 – Connecting to a data source

2. You can search for an existing data source that has been defined, or you can add a new data connection. If you select the **add new data connection** option (shown in the preceding figure), you will see the following connection choices:

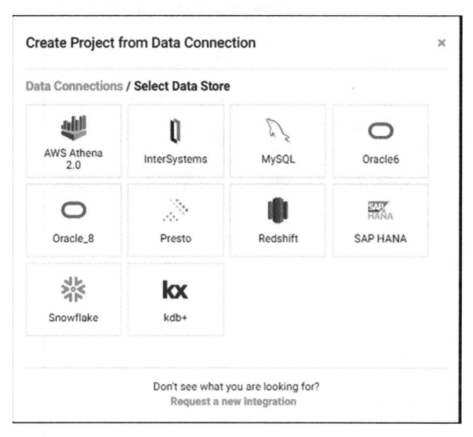

Figure 4.2 – Types of data connection

3. You will see the connection choices available for your organization. What you see here could be different from the preceding figure. Most databases with JDBC drivers are supported, but you might have to check with your administrator. As an example, let's select the **MySQL** option, as shown in the following figure:

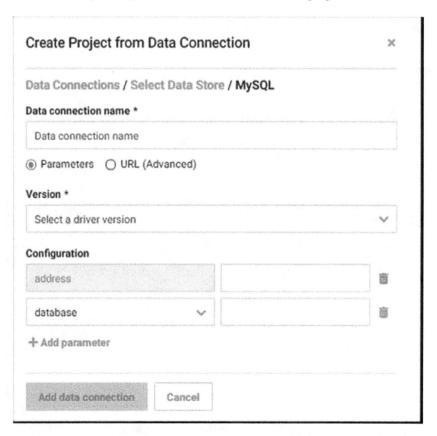

Figure 4.3 – Configuring a data connection

In the preceding figure, you will see the configuration parameters for configuring a MySQL data source. Other data sources are similar in nature. Here, you will enter the configuration settings that can be obtained from your database administrator. You will need to create a similar connection if you are connecting to a database to get data into Python or Excel.

> **Note**
> You will need to have some working knowledge of SQL or work with someone who knows SQL to make use of these options.

Aggregating data for modeling

From the previous chapters, you might remember that machine learning algorithms expect the dataset to be in a specific form and it needs to be in one table. The data needed for this table, however, could reside in multiple sources. Hence, one of the first things you need to do is to aggregate data from multiple sources. This is often done using SQL or Python. Recently, DataRobot has added the capability to add multiple datasets into a project and then aggregate this data within DataRobot. Please note that there are still some data cleansing operations that you might have to do outside of DataRobot, so if you want to use the aggregation capabilities of DataRobot, you need to do cleansing operations prior to bringing this data into DataRobot. We cover data cleansing in the following section. If you choose to do data aggregation inside DataRobot, you have to make sure to do this at the very start of the project (*Figure 4.4*):

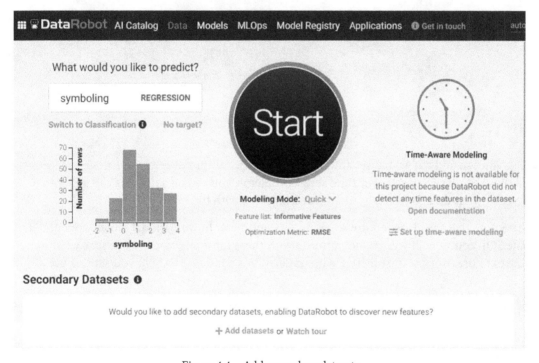

Figure 4.4 – Add secondary datasets

In the preceding figure, just below the **Start** button, you can click on **Add datasets**. Once you click on it, you will see a window that lets you specify the additional dataset, as shown in *Figure 4.5*:

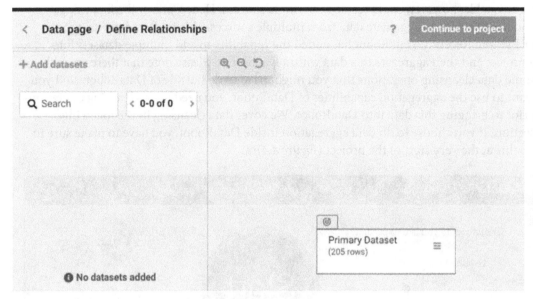

Figure 4.5 – Secondary datasets

Here, you can add a new dataset and define the relationships between your main dataset and the secondary datasets. For time series problems, you can also use this capability to aggregate your data to the right timescale and join it with the main dataset.

Please note that this does require some understanding of how relational tables work and some SQL concepts. If you are not familiar with these ideas and you are not sure what indexes to use, work with someone who understands databases to help you set this up.

Cleansing the dataset

This step can come before or after the data aggregation we talked about in the previous section. We introduced some concepts around data cleansing in *Chapter 2, Machine Learning Basics*, so let's look at how to actually do it on a dataset. For this, let's start with the Automobile Dataset. Please refer to the *Technical requirements* section to access the UCI repository for this dataset:

1. Let's download two files: `imports-85.data` and `imports-85.names`. The data file is in `.csv` format, so let's rename the file with the `.csv` extension and open it using Excel (you can use any text editor). You will now see the data (*Figure 4.6*):

	A	B	C	D	E	F	G	H	I	J	K	L	M	N	O	P	Q	R
1	3	?	alfa-rome	gas	std	two	convertibl	rwd	front	88.6	168.8	64.1	48.8	2548	dohc	four		130 mpfi
2	3	?	alfa-rome	gas	std	two	convertibl	rwd	front	88.6	168.8	64.1	48.8	2548	dohc	four		130 mpfi
3	1	?	alfa-rome	gas	std	two	hatchback	rwd	front	94.5	171.2	65.5	52.4	2823	ohcv	six		152 mpfi
4	2	164	audi	gas	std	four	sedan	fwd	front	99.8	176.6	66.2	54.3	2337	ohc	four		109 mpfi
5	2	164	audi	gas	std	four	sedan	4wd	front	99.4	176.6	66.4	54.3	2824	ohc	five		136 mpfi
6	2	?	audi	gas	std	two	sedan	fwd	front	99.8	177.3	66.3	53.1	2507	ohc	five		136 mpfi
7	1	158	audi	gas	std	four	sedan	fwd	front	105.8	192.7	71.4	55.7	2844	ohc	five		136 mpfi
8	1	?	audi	gas	std	four	wagon	fwd	front	105.8	192.7	71.4	55.7	2954	ohc	five		136 mpfi

Figure 4.6 – Automobile data

2. You will notice in the preceding screenshot that it is missing the header information. To retrieve the header information, open the `.names` file in any text editor. You will see the names of attributes as well as their definitions. Create an empty row at the top of your `.csv` file and you will have to manually type the names of these attributes as the first row of your file. Now let's save this file as `autodata.csv`. It should now look as shown in *Figure 4.7*:

	A	B	C	D	E	F	G	H	I	J	K	L
1	symboling	normalized-losses	make	fuel-type	aspiration	num-of-doors	body-style	drive-whe	engine-lo	wheel-ba	length	width
2	3	?	alfa-romero	gas	std	two	convertible	rwd	front	88.6	168.8	64.1
3	3	?	alfa-romero	gas	std	two	convertible	rwd	front	88.6	168.8	64.1
4	1	?	alfa-romero	gas	std	two	hatchback	rwd	front	94.5	171.2	65.5
5	2	164	audi	gas	std	four	sedan	fwd	front	99.8	176.6	66.2
6	2	164	audi	gas	std	four	sedan	4wd	front	99.4	176.6	66.4
7	2	?	audi	gas	std	two	sedan	fwd	front	99.8	177.3	66.3

Figure 4.7 – Automobile data with headers

Please review all the cells in this data file. You will have already noticed that many cells in the preceding figure have a **?** instead of a value. While there are several features where the values are missing, for most of them it is negligible, except for `normalized-losses` where 20% of the total values are missing. Given that our dataset is very small, we do not want to drop the rows with missing data. Also, DataRobot has mechanisms to account for missing values, so we are going to leave most of them as is. The only one that we want to consider is `normalized-losses`. If `normalized-losses` is our target variable, then we have no choice but to drop those rows. If not, we can first try to go as is and let DataRobot build a model. We can then try an alternative strategy of using the average value of `normalized-losses` per **Symboling** value to see if that makes any difference. I will use Excel's pivot table functionality to compute these averages (*Figure 4.8*):

Symboling	
Row Labels	Average of normalized-losses
-2	103
-1	85.6
0	113.1666667
1	128.5744681
2	125.6896552
3	168.6470588

Figure 4.8 – Pivot table

The reason for using **Symboling** is that it is an indicator of risk. Depending on the problem and what you are trying to accomplish, you can choose some other feature for this purpose. For now, we will use **Symboling** to illustrate how to do it. There are more sophisticated imputation methods available, such as a K-Nearest Neighbor-based imputation method, that you can explore if desired (`https://scikit-learn.org/stable/modules/generated/sklearn.impute.KNNImputer.html`).

In reviewing the Appliances Energy Prediction Dataset, we see that the data looks very clean and no further cleansing is required. In real-world projects, you will almost never find a dataset that is free of problems. Typical problems in time series datasets to watch out for are as follows:

- **Very little data**: You need at least 35 or so datapoints for regression and 100 datapoints for classification problems to allow DataRobot to do something useful with your data.

- **Data gaps**: Sometimes data might be missing for certain timesteps. In these cases, you can use values from the timesteps before or after to assign values for the missing time step. You can also let DataRobot do this for you.

- **Interrelated series**: Often you will have multiple timeseries that you are trying to forecast. If the series are similar and are interrelated, then you can combine them into a single model. This can often improve the forecast accuracy. In these cases, you have to create a feature that tells DataRobot that these series are part of the same cluster.

We will revisit the data quality based on what DataRobot finds. Now that the dataset looks reasonably clean (which is very unusual by the way), let's investigate this data further.

Working with different types of data

You will have noticed that some of the features have numeric values while others have categorical values. For example, the **aspiration** feature can have two values: **std** or **turbo**. Such categorical features require some preprocessing to convert them into numeric values. Luckily DataRobot takes care of that processing for us. You might want to check for misspellings, though, to make sure that the possible values match the expectations. For example, you might find `standard` as well as `std` in your datasets. In this case, DataRobot will treat them as different values, even though they are the same.

There are some features that can be treated as categorical or as numerical. For example, **Symboling** can be treated as numerical or as categorical. In general, if the numerical value has meaning, it is better to treat it as numerical. Another example is `num-of-cylinders`; here, the values are expressed as text. Given that there is a numerical order here, it might be beneficial to turn this into a numeric variable, as shown in *Figure 4.9*:

	A	B	C	D	E	F	G	H	I	J	K	L	M	N	O	P	Q
	symboling	normalized-losses	make	fuel-type	aspiration	num-of-doors	body-style	drive-wheels	engine-location	wheel-base	length	width	height	curb-weight	engine-type	num-of-cylinders	cylinder_count
1																	
2	3	168.6470588	alfa-rome	gas	std	two	convertibl	rwd	front	88.6	168.8	64.1	48.8	2548	dohc	four	4
3	3	168.6470588	alfa-rome	gas	std	two	convertibl	rwd	front	88.6	168.8	64.1	48.8	2548	dohc	four	4
4	1	128.5744681	alfa-rome	gas	std	two	hatchback	rwd	front	94.5	171.2	65.5	52.4	2823	ohcv	six	6
5	2	164	audi	gas	std	four	sedan	fwd	front	99.8	176.6	66.2	54.3	2337	ohc	four	4
6	2	164	audi	gas	std	four	sedan	4wd	front	99.4	176.6	66.4	54.3	2824	ohc	five	5
7	2	125.6896552	audi	gas	std	two	sedan	fwd	front	99.8	177.3	66.3	53.1	2507	ohc	five	5
8	1	158	audi	gas	std	four	sedan	fwd	front	105.8	192.7	71.4	55.7	2844	ohc	five	5

Figure 4.9 – Categorical to numerical feature conversion

Here, we have created (in **Excel**) a new feature, `cylinder-count`, that carries the numerical values for the number of cylinders. In this example, we are using Excel for the data manipulation, but this can be achieved via many methods, such as SQL, Python, and Paxata. You can do similar data manipulation and create a new column for `num-of-doors` as well.

Let's take a look at the `make` feature in the following figure. This seems to have 22 possible values, but we have very limited data available. If we count the number of rows for each make, we can see how much data is available for each make:

make	Count
mercury	1
renault	2
alfa-romero	3
chevrolet	3
jaguar	3
isuzu	4
porsche	5
saab	6
audi	7
plymouth	7
bmw	8
mercedes-benz	8
dodge	9
peugot	11
volvo	11
subaru	12
volkswagen	12
honda	13
mitsubishi	13
mazda	17
nissan	18
toyota	32

Figure 4.10 – Data for each make

We notice that some car types have very little data available, so it might be useful to combine some of them. For example, we can combine (using Excel) the highlighted rows into a make called other. Where you draw the line depends upon your understanding of the business problem or discussions with domain experts. Even with that knowledge, you might have to try out a few different options to see what works best. This is what makes machine learning an iterative and exploratory process. Also keep in mind that you have limited time available, so don't over-explore either. There is certainly a point of diminishing returns where additional tinkering will not produce many benefits.

DataRobot also allows special processing for images and geo-spatial data. We will cover them in *Chapter 11, Working with GeoSpatial Data, NLP, and Image Processing.* Now let's look at other transformations that can be done on data.

Engineering features for modeling

As part of the system's understanding you would have gained some insights into your problem and dataset that can be used to create new features in your dataset by combining the existing features in various ways. For example, we can create a new feature called volume by multiplying length, width, and height. Similarly, we can create a feature called mpg-ratio by dividing highway-mpg by city-mpg. Let's also create a feature called cylinder-size by dividing engine-size by cylinder-count. The equations for these features are as follows:

- volume = length * width * height

- mpg-ratio = highway-mpg / city-mpg

- cylinder-size = engine-size / cylinder-count

Figure 4.11 shows an example of what these feature values look like:

	L	M	N	O	P	Q	R	S	T	U	V	W	X	Y	Z	AA	AB	AC	AD
	width	height	curb-weight	engine-type	num-of-cylinders	cylinder-count	engine-size	fuel-system	bore	stroke	compres-sion-ratio	horsepo-wer	peak-rpm	city-mpg	highway-mpg	price	volume	mpg-ratio	cylinder-size
1																			
2	64.1	48.8	2548	dohc	four	4	130	mpfi	3.47	2.68	9	111	5000	21	27	13495	528019.9	1.285714	32.5
3	64.1	48.8	2548	dohc	four	4	130	mpfi	3.47	2.68	9	111	5000	21	27	16500	528019.9	1.285714	32.5
4	65.5	52.4	2823	ohcv	six	6	152	mpfi	2.68	3.47	9	154	5000	19	26	16500	587592.6	1.368421	25.33333
5	66.2	54.3	2337	ohc	four	4	109	mpfi	3.19	3.4	10	102	5500	24	30	13950	634817	1.25	27.25
6	66.4	54.3	2824	ohc	five	5	136	mpfi	3.19	3.4	8	115	5500	18	22	17450	636734.8	1.222222	27.2
7	66.3	53.1	2507	ohc	five	5	136	mpfi	3.19	3.4	8.5	110	5500	19	25	15250	624190	1.315789	27.2
8	71.4	55.7	2844	ohc	five	5	136	mpfi	3.19	3.4	8.5	110	5500	19	25	17710	766364	1.315789	27.2
9	71.4	55.7	2954	ohc	five	5	136	mpfi	3.19	3.4	8.5	110	5500	19	25	18920	766364	1.315789	27.2

Figure 4.11 – Engineered features for the Automobile Dataset

As you can now see, many possibilities exist to create new features that could prove helpful in solving your problem. Many of these new features may not be useful, and it is OK to drop them later. Sometimes, such features will have meaning for the customers or stakeholders, and you might want to keep them instead of some other features that are redundant.

Let's take a look at the Appliances Energy Prediction Dataset file. With this dataset, we can create the following features:

- `total-energy = Appliances + lights`

- `avg-temp-inside = (T1 + T2 + T3 + T4 + T5 + T7 + T8 + T9 ) / 8`

- `avg-rh-inside = (RH_1 + RH_2 + RH_3 + RH_4 + RH_5 + RH_7 + RH_8 + RH_9 ) / 8`

- `temp-inout-diff = T6  -  avg-temp-inside`

- `rh-inout-diff = RH_6  -  avg-rh-inside`

- `windchill-factor` (I am creating an approximate windchill factor based on https://www.weather.gov/media/epz/wxcalc/windChill.pdf) `= T_out * (Windspeed0.16 )`

The new data features will appear as shown in *Figure 4.12*:

	A	T	U	V	W	X	Y	Z	AA	AB	AC	AD	AE	AF	AG	AH	AI
	date	T9	RH_9	T_out	Press_mm_hg	RH_out	Windspeed	Visibility	Tdewpoint	rv1	rv2	total-energy	avg-temp-inside	avg-rh-inside	temp-inout-diff	rh-inout-diff	windchill-factor
1																	
2	1/11/2016 17:00	17.03333	45.53	6.60E+00	733.5	92	7	63	5.30E+00	13.27543	13.27543	90	18.435	46.7425	-11.40833	37.51417	9.01E+00
3	1/11/2016 17:10	17.06667	45.56	6.48E+00	733.6	92	6.66666667	59.16667	5.20E+00	18.60619	18.60619	90	18.43917	46.67271	-11.60583	37.39063	8.78E+00
4	1/11/2016 17:20	17	45.5	6.37E+00	733.7	92	6.33333333	55.33333	5.10E+00	28.64267	28.64267	80	18.42167	46.56292	-11.86167	36.59375	8.55E+00
5	1/11/2016 17:30	17	45.4	6.25E+00	733.8	92	6	51.5	5.00E+00	45.41039	45.41039	90	18.39625	46.46875	-11.96292	36.95458	8.32E+00
6	1/11/2016 17:40	17	45.4	6.13E+00	733.9	92	5.66666667	47.66667	4.90E+00	10.0841	10.0841	100	18.40875	46.46292	-12.04208	38.43042	8.10E+00
7	1/11/2016 17:50	17	45.29	6.02E+00	734	92	5.33333333	43.83333	4.80E+00	44.91948	44.91948	90	18.39208	46.42	-12.09208	39.34667	7.86E+00
8	1/11/2016 18:00	17	45.29	5.90E+00	734.1	92	5	40	4.70E+00	47.23376	47.23376	110	18.38792	46.37542	-12.12458	39.71458	7.63E+00
9	1/11/2016 18:10	17	45.29	5.92E+00	734.16667	91.83333	5.16666667	40	4.68E+00	33.03989	33.03989	110	18.37208	46.35042	-12.18208	40.07292	7.69E+00
10	1/11/2016 18:20	17	45.29	5.93E+00	734.23333	91.66667	5.33333333	40	4.67E+00	31.4557	31.4557	100	18.38042	46.36135	-12.25708	40.86531	7.76E+00
11	1/11/2016 18:30	17	45.29	5.95E+00	734.3	91.5	5.5	40	4.65E+00	3.089314	3.089314	110	18.39583	46.47875	-12.20583	41.14792	7.82E+00
12	1/11/2016 18:40	17	45.29	5.97E+00	734.36667	91.33333	5.66666667	40	4.63E+00	10.29873	10.29873	300	18.42635	46.64885	-12.23635	41.21781	7.88E+00

Figure 4.12 – Engineered features for the Appliances Energy Prediction Dataset

As you can see, these features use our knowledge about the domain that we can find by talking to domain experts or doing some research on the internet. You might be able to find even more such features by doing some research about dew points, pressure, and visibility. It will be hard for the automation to catch all of these on its own, but on the other hand, the automation might be able to find some additional interesting features based on them. Recently, DataRobot has also been adding capabilities to automatically do some feature engineering, but these capabilities are somewhat limited. One area where these capabilities are very useful is time series problems. In this particular area, these capabilities are extremely helpful in trying out a wide range of features that will be hard to match on your own. Having said that, it is still your responsibility to inject your domain knowledge into the model via engineered features.

Summary

In this chapter, we covered methods to help you prepare the dataset for building the models. Many of these methods have to be applied outside of DataRobot, although DataRobot is beginning to provide support for many of the data preparation tasks. As we discussed, many of these tasks cannot be automated at this point in time, and they require domain understanding to make appropriate decisions.

Specifically, in this chapter we have learned how to connect to various data sources and how to aggregate data from these sources. We looked at examples to address missing data issues and other data manipulation that should be done prior to modeling. We also covered several methods for creating new features that can be very important for improving the model's performance.

We are now at a stage where we will be working almost completely inside the DataRobot environment to analyze the data and build models. In the next chapter, we will use DataRobot to analyze the datasets.

5
Exploratory Data Analysis with DataRobot

In this chapter, we will cover tasks related to exploring and analyzing your dataset with DataRobot. DataRobot performs many functions that you will need to perform this analysis, but it is still up to you to make sense of it.

By the end of this chapter, you will have learned how to utilize DataRobot to perform **exploratory data analysis (EDA)**. In this chapter, we're going to cover the following main topics:

- Data ingestion and data cataloging
- Data quality assessment
- EDA
- Setting the target feature and correlation analysis
- Feature selection

Data ingestion and data cataloging

Now that we have our datasets ready, we have two choices to bring them into DataRobot. We can go to either the **Create New Project / Drag Dataset** page (*Figure 1.5*) or the **AI Catalog** page (*Figure 1.17*). If the dataset is relatively small, we may prefer to start with the **Create New Project** method. After a few iterations, when the dataset has stabilized, you can move it into the **AI Catalog** page so that it can be reused in other projects.

Let's start by uploading our automobile dataset as a local file that we created in *Chapter 4, Preparing Data for DataRobot*. You can name the project Automobile Example 1, as shown in the following screenshot:

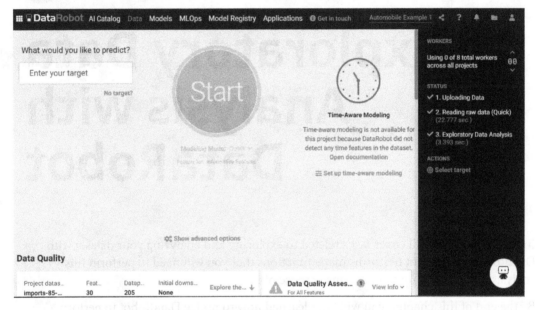

Figure 5.1 – Uploading dataset for a new project

You will notice that DataRobot automatically starts analyzing the data and performs a quick exploratory analysis. You can see that it found 30 features and 205 rows of data.

> **Note**
> If you are using an Excel file that has multiple sheets, make sure that the data you want is in the first sheet.

Data quality assessment

DataRobot will also perform a data quality assessment and notify you if it finds any data issues, as shown in the following screenshot:

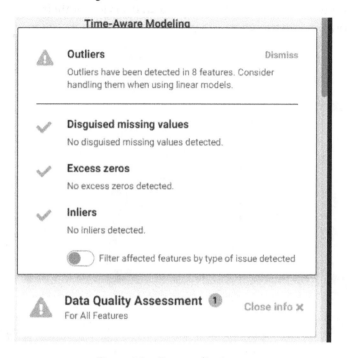

Figure 5.2 – Data quality issues

In this case, it has found outliers in eight features. You can look into the details to see if these look acceptable or if you need to drop or otherwise fix these outliers. We will do this as we explore and analyze each of these features in the following section.

Notice that it also looked for any disguised missing values or excess zeros in any feature. These can be hard to detect manually and can be problematic for your models, so it is important to fix these issues if they come up. For example, you saw in *Chapter 4, Preparing Data for DataRobot,* that we already fixed the issue of excess zeros in the normalized-losses feature. If we had not done that previously, DataRobot would alert us to fix this or filter out those rows before proceeding. It will also perform additional analysis once a target feature is selected.

You will carry out the same process with the Appliances Energy dataset.

EDA

As you saw in the previous section, DataRobot automatically performed an initial analysis of the dataset. Let's see how we will review this data and gain insights from it. If you scroll down the page, you will see a table of features and an overview of their characteristics, as shown in the following screenshot:

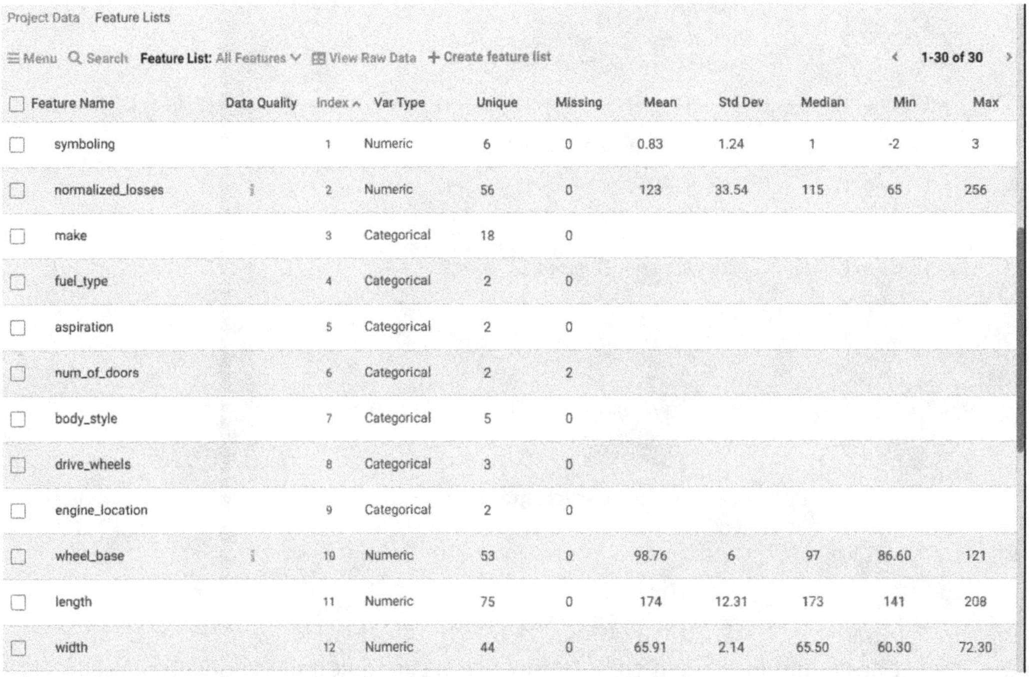

Feature Name	Data Quality	Index ∧	Var Type	Unique	Missing	Mean	Std Dev	Median	Min	Max
symboling		1	Numeric	6	0	0.83	1.24	1	-2	3
normalized_losses	i	2	Numeric	56	0	123	33.54	115	65	256
make		3	Categorical	18	0					
fuel_type		4	Categorical	2	0					
aspiration		5	Categorical	2	0					
num_of_doors		6	Categorical	2	2					
body_style		7	Categorical	5	0					
drive_wheels		8	Categorical	3	0					
engine_location		9	Categorical	2	0					
wheel_base	i	10	Numeric	53	0	98.76	6	97	86.60	121
length		11	Numeric	75	0	174	12.31	173	141	208
width		12	Numeric	44	0	65.91	2.14	65.50	60.30	72.30

Figure 5.3 – Data analysis overview

You can see that in this table, DataRobot has computed and listed any data quality concerns regarding a feature, what type of variable it is, how many unique values are in the dataset, and how many values are missing. These are all very important characteristics, and you need to review all of them to make sure that you understand what they are telling you.

For example, is the variable type selected by DataRobot what you expected? If you look at num_of_doors, you will notice that this is categorical. Even though this is correct because the data contained is in the form of text, you know that this is really numbers. You might want to fix this (just as we did for num_of_cylinders in *Chapter 4, Preparing Data for DataRobot*). Doing this ahead of time will reduce rework and wasted effort downstream. Similarly, you will notice that num_of_doors has two missing values. If this number were higher, we would have tried to address the missing values, as discussed in *Chapter 4, Preparing Data for DataRobot*. Also, pay attention to unique values. For some features, we expect many unique values, while for others, we do not. Check if what DataRobot found is consistent with your expectations. If not, try to determine the reason for this. Pay special attention when a categorical variable has a large number of unique values. We will soon discuss how to address this issue.

For numeric features, you will also see summary statistics such as **Mean, Median, Std Dev** (for standard deviation), **Min**, and **Max**. Review these for each feature to see if they all look reasonable. If you click on any feature row, it will expand and show more detail, as shown in the following screenshot:

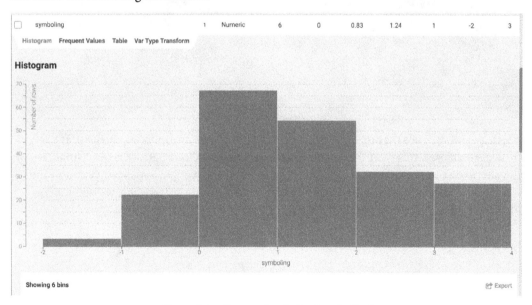

Figure 5.4 – Feature details for "symboling"

Here, you can see a histogram of all the values. You can now see how this data is distributed. One aspect to pay special attention to is the area where you don't have much data. For example, you can see that the amount of training data available for the value -2 is very limited, so we should expect there to be problems trying to predict these values. Now, let's look at the details of `normalized_losses` in the following screenshot:

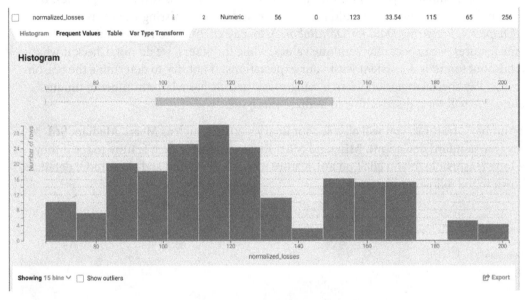

Figure 5.5 – Feature details for "normalized_losses"

In this view, we can see that there seem to be very few losses around **140** and **180**. If this were a large dataset, this would be a cause for concern. Since our dataset is very small, it is not surprising to see such gaps. Also, note that these are average losses per year and not losses experienced by an individual car. Next, in the following screenshot, we will look at the `make` feature to see how it is distributed:

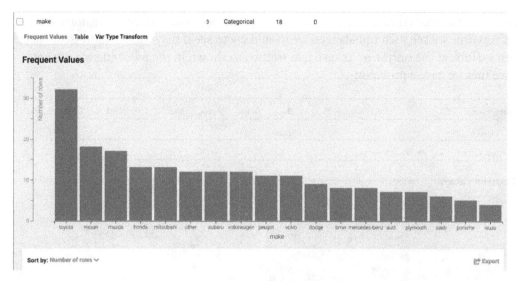

Figure 5.6 – Feature details for "make"

Since make is a categorical feature, you can see how frequently each value shows up. Remember that we had already consolidated some car types that had very little data into other. If we hadn't done that, we would notice here that some types have very few data points and need to be addressed or they will not do well during training. Let's look at fuel_type to see what we can glean from this data, as shown in the following screenshot:

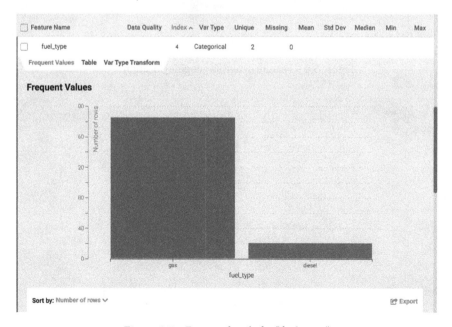

Figure 5.7 – Feature details for "fuel_type"

Here, we notice that `diesel` cars are not well represented, and this might be normal for cars. Anytime we see such imbalances, we should try to see if they can be addressed. Now, when we look at the `engine_location` feature, as shown in the following screenshot, we see that we have a problem:

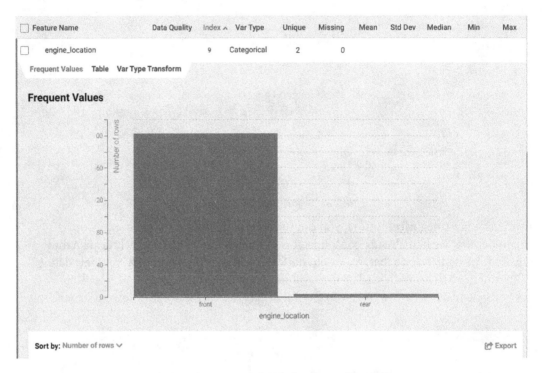

Figure 5.8 – Feature details for engine_location

As you can see in the preceding screenshot, the `rear` feature is barely registering on the dataset. From a practical standpoint, what this means is that the algorithms will ignore this feature. If you did not look carefully, you might assume that `engine_location` has no impact on your target, but as you can tell from this screenshot, our dataset is not large enough to make that determination. Let's now look in the following screenshot at `engine_type` to see what we find here:

Figure 5.9 – Feature details for "engine_type"

In this case, we find that one type dominates and some of the types are barely represented. Looking at this distribution, you might want to create another feature where you transform this into a binary value, **0** for `ohc` and **1** for every other type. This will also create some balance in the dataset.

Please bear in mind that this might or might not prove to be useful. You have to try it out in your models and see what works. Let's now look in the following screenshot at `num_of_cylinders` and `cylinder_count`, a feature that we created during data preparation:

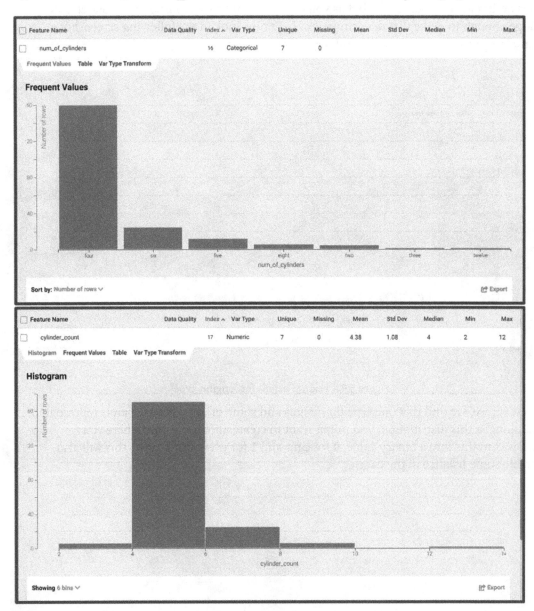

Figure 5.10 – Feature details for "num_of_cylinders" and "cylinder_count"

As you can see, even though it is the same data, transforming the values provides a different impression compared to what you get when you first look at the histograms. The numeric values are a more accurate representation of the data and should result in a better model compared to the categorical values.

Hopefully, we have highlighted what DataRobot provides automatically and what kinds of insights can be gained by looking at the graphs generated by DataRobot. We are now ready to set our target feature and do additional analysis.

Setting the target feature and correlation analysis

By the time you reach this stage, you should already have a pretty good idea of the problem you are trying to solve and what should be the target feature. It is not unusual to use different features as targets for different use cases. Also, sometimes you will set a transformed feature as a target (for example, log of a feature). For the Automobile dataset, we want to predict the **price** of cars. Once you select the target feature, as shown in the following screenshot, it will analyze that feature and provide some recommendations:

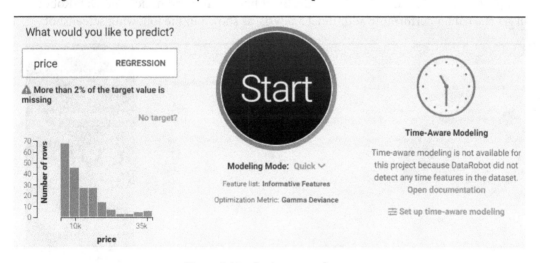

Figure 5.11 – Setting target feature

You can see from the preceding screenshot that it is showing how the price is distributed. DataRobot also cautions that some of the target values are missing. Ideally, we would filter out the rows with missing target values before uploading the dataset. You will also notice that DataRobot has characterized this as a regression problem. Another thing to note is that it has picked the optimization metric to be **Gamma Deviance**. You can read more about this metric in *Chapter 2, Machine Learning Basics*, or you can explore it in more detail in DataRobot's help sections. For now, it looks like a good choice, given the wide variance of price values.

Before we click on the **Start** button, we should explore the advanced options. The reason for this is that once you click the **Start** button, you cannot make changes to the options. Having said that, it is often hard to make all the right choices without completely understanding the data. One way to overcome this issue is to ignore the advanced options for now and go ahead with the exploration.

Once we know what we want, we can create a new project and select the appropriate options. You can see that this is an iterative process, and we will often try something and come back and redo some of it. Also, notice that **Modeling Mode** in *Figure 5.11* is set to **Quick**. This is normally a good choice to get started. With that in mind, we can actually skip the options and go ahead and click the **Start** button. You will notice that DataRobot will get started on performing additional analysis, as shown in the following screenshot:

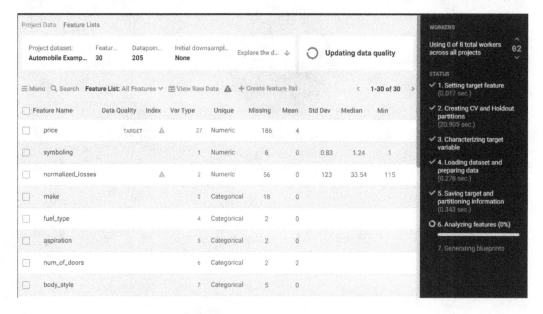

Figure 5.12 – Feature analysis

You will notice that in addition to performing additional analysis, DataRobot will actually start building the models. This might be surprising since we are still doing analysis, but fear not—these are not the final models. Let DataRobot build these models, as some of these will provide useful insights into our data. We will most likely discard these models later on, but they will prove useful in our journey. Once DataRobot has finished doing all the tasks, you will see an **Autopilot has finished** message, as shown in the following screenshot:

Figure 5.13 – Initial analysis complete

You will now notice that DataRobot has populated an **Importance** column for all the features. This is the relative importance of a feature in reference to the target feature. We can also check to see if there are additional data quality issues that have been found. For that, let's click on the **View info** dropdown in the **Data Quality Assessment** box. You will then see the options, as shown in the following screenshot:

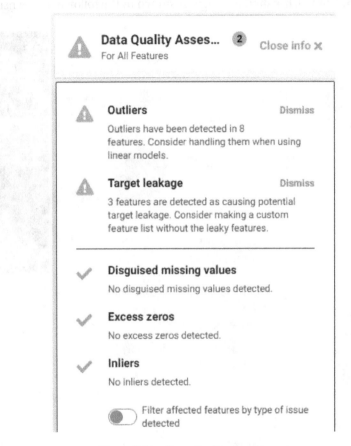

Figure 5.14 – Data Quality Assessment

We saw some of the issues previously, but we now see that there are features that potentially have target leakage. If target leakage exists, we will filter those features out. By looking at the warning signs associated with each feature, we discover that these features are horsepower and engine_size. Since these are important features and have an obvious impact on price, we will retain these features. We also see another warning symbol in the header row, as shown in the following screenshot:

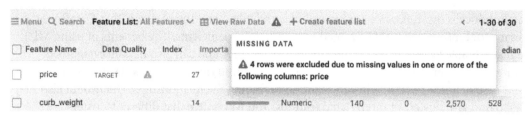

Figure 5.15 – Missing target values

Clicking on the symbol, we see that DataRobot has already filtered out rows where the price is missing. This is good, as it means we don't have to recreate our dataset and upload it again into DataRobot. You will also notice in the following screenshot that a new tab called **Feature Associations** is now present at the top left of the screen. This is a critical tab for our data analysis task. Let's click on this tab to look at what DataRobot has found:

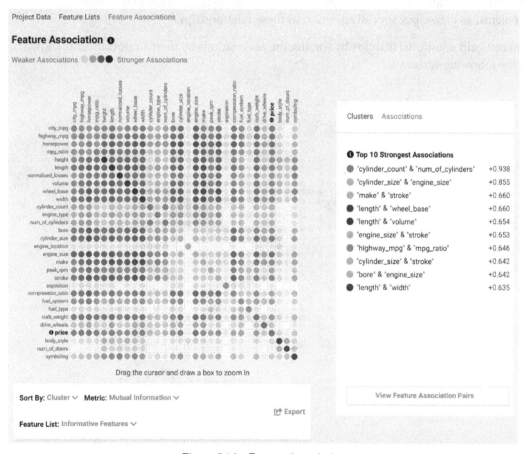

Figure 5.16 – Feature Association

DataRobot calls these *associations* instead of *correlations*, and the reason is that DataRobot uses **mutual information** (**MI**) instead of correlation coefficients. The benefits of using MI are that it is able to better reflect non-linear relationships and can also handle categorical features. If you perform correlation analysis, you will find that the results are very similar in the case of linear relationships. In addition to finding the relationships, DataRobot also tries to find any clusters of interrelated features. You will notice that different clusters are color-coded differently. These offer additional insights into your problem. For example, you will see that there is a cluster of features related to the `engine` that includes a group of tightly correlated features such as `engine_size`, `bore`, `cylinder_size`, and `stroke`. Understanding these relationships as a collective can be very important to solving a business problem. In this particular case, it tells you that you cannot modify one of these in isolation.

Changing the bore will affect many other features, even if your model does not end up with those features. Ignoring these aspects is what typically leads to downstream problems, so please pay special attention to these relationships.

You can gain additional insights by sorting the associations by their importance, as shown in the following screenshot:

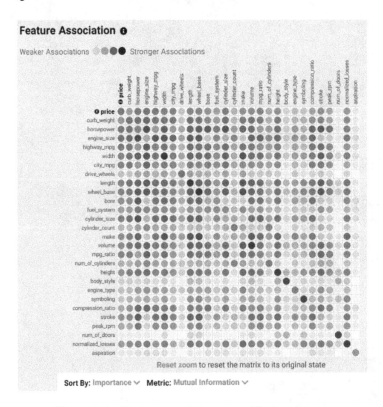

Figure 5.17 – Feature associations sorted by importance

The preceding screenshot shows the features sorted by their impact on the target feature. This tells you which features are most likely to be prominent in your model. One thing to look for is how this lines up with the causal model that you built during the problem understanding stage. Is it consistent? If not, where are the differences and surprises? These typically lead to new insights into your problem. It is also useful to look at the MI values in totality. For this, you can click on the **Export** button to export all MI values as a `.csv` file. You can then analyze them in tools such as Excel, as shown in the following screenshot:

feature1	feature2	statistic
price	engine_size	0.47
price	horsepower	0.45
price	curb_weight	0.45
price	highway_mpg	0.44
price	cylinder_size	0.43
price	mpg_ratio	0.42
price	city_mpg	0.41
price	width	0.41
price	length	0.41
price	wheel_base	0.41
price	volume	0.38
price	stroke	0.37
price	make	0.37
price	height	0.37
price	bore	0.35
price	normalized_losses	0.34
price	fuel_system	0.31
price	peak_rpm	0.29
price	compression_ratio	0.29
price	num_of_cylinders	0.25
price	drive_wheels	0.24
price	cylinder_count	0.24
price	engine_type	0.21
price	symboling	0.19
price	body_style	0.13
price	aspiration	0.09
price	num_of_doors	0.07

Figure 5.18 – MI values

This gives you a better feel for the relative scale of these values. In this view, we can see that aspiration has very little impact on price. This seems a little counterintuitive and merits some additional investigation. For this, we can look at this association in more detail by clicking on the **View Feature Association Pairs** button. You can now select price and aspiration to see the association details, as shown in the following screenshot:

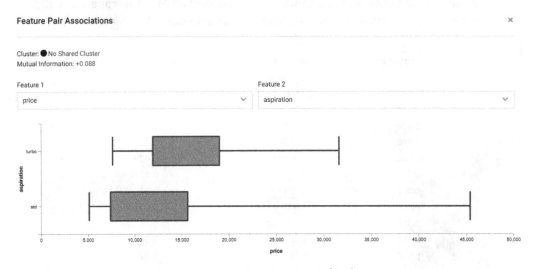

Figure 5.19 – Association pair details

Here, we can see that for the same value of aspiration, the price can vary quite a bit. Still, we can see that on average, turbo has a higher price. Based on this, we will keep it in the mix for modeling. We should also discuss with the domain experts to see why it is not correlating in a stronger fashion with price. These discussions can lead to creating other features that might clarify this relationship. On the other hand, the relationship between price and num_of_doors doesn't look very interesting.

It is a good idea to review the association pairs to see what insights can be gained. At a minimum, review the ones with very high or very low values. Specifically, look for non-linear relationships. For example, let's look at the association between curb_weight and highway_mpg, as shown in the following screenshot:

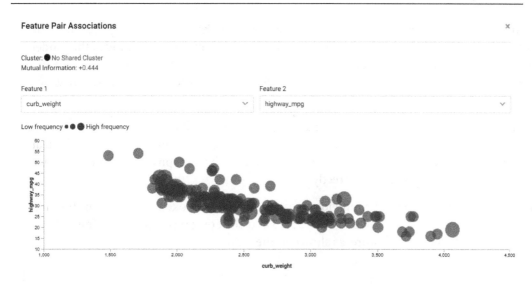

Figure 5.20 – Association between curb_weight and highway_mpg

Here, you will notice that as `curb_weight` increases, the **miles per gallon (MPG)** value decreases, which makes intuitive sense. We also see that the curve starts flattening at higher weights. This could be due to many reasons, as other factors affecting MPG do not increase with weight.

Note that while this may or may not affect the predictive accuracy of the model, understanding these relationships is key to determining actions to be taken based on the model. For example, weight reduction might not provide much MPG benefit for weights larger than **3500**. We can also investigate the association between `curb_weight` and `drive _wheels`, as shown in the following screenshot:

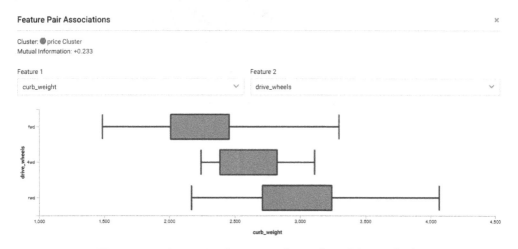

Figure 5.21 – Association between curb_weight and drive_wheels

In the preceding screenshot, we can see that `curb_weight` is impacted by the choice of `drive_wheels`. It is possible that if we use both these features in our model, the model will give a much higher preference to `curb_weight` and might find not much value in using `drive_wheels`. Business users might therefore interpret `drive_wheels` as not very important.

As you can see, this is not true since `curb_weight` is itself influenced by `drive_wheels`. It has been observed that an accurate model can sometimes give a false impression if you are not careful. DataRobot can do this analysis and produce these graphs, but it is up to you to understand and interpret these correctly.

Let's look again at some of the individual feature graphs we looked at before. For this, let's look at the feature details shown in *Figure 5.13* and click on `curb_weight`. This will show us details about the feature, as shown in the following screenshot:

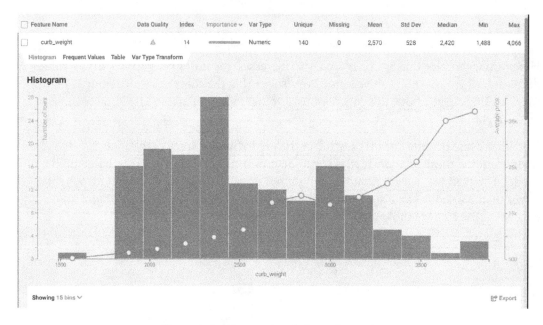

Figure 5.22 – Feature details for curb_weight

You will notice that we now have some more information in this graph. Specifically, we can now see how price varies with `curb_weight` as well as how the `curb_weight` value is distributed. Looking at these relationships can give you additional insights into your problem, especially when the relationship is non-linear. For example, let's look at the details for `highway_mpg` in the following screenshot:

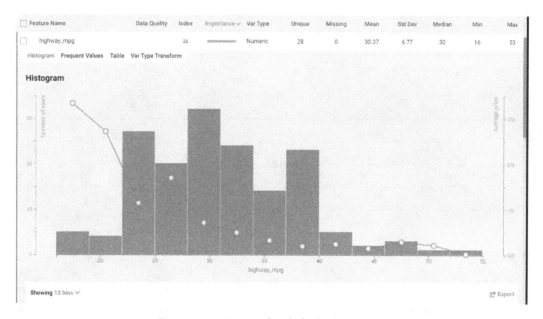

Figure 5.23 – Feature details for highway_mpg

As you can see, the price drops exponentially as the MPG value increases. Given this non-linearity, which also seems to be present in other features, it might be useful to try creating a new target feature by taking a log of the price. Similarly, by looking at the other features, you can get ideas on feature transformations that might prove beneficial. Some of you might be wondering why we should do this since the new algorithms can handle non-linearity. While that is true, it is still better to transform your non-linear problems if it makes sense from a business-understanding perspective. Also, it allows the algorithm to focus its computational energy in other areas that might otherwise be overlooked.

Now that we have understood the features and have transformed them as needed, we can focus on selecting a feature set to start the modeling process.

Feature selection

The basic idea behind feature selection is to select features that show high importance for the target. In addition, we want to remove any features that are highly cross-correlated (or have high MI values) to other features. The selected set of features are represented as feature lists in DataRobot. If you click on the **Feature Lists** menu on the top left of the page, as shown in the following screenshot, you will see the feature lists that DataRobot has created for the dataset:

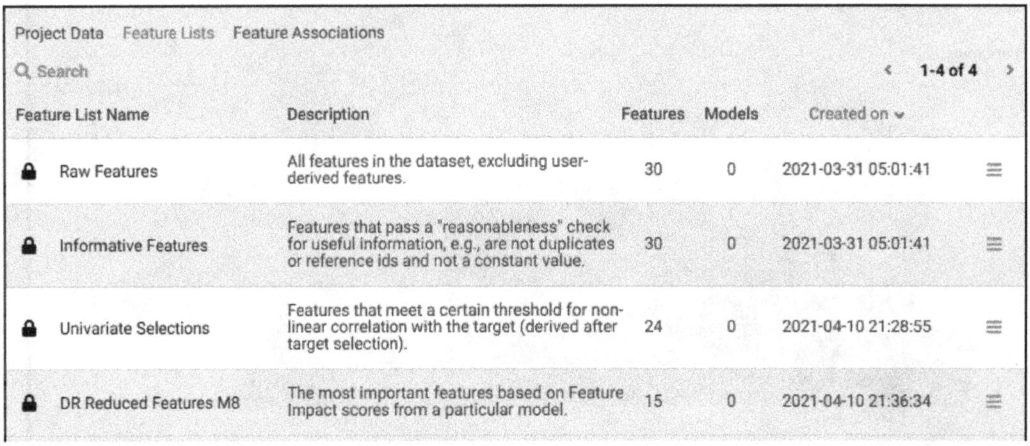

Feature List Name	Description	Features	Models	Created on ⌄	
🔒 Raw Features	All features in the dataset, excluding user-derived features.	30	0	2021-03-31 05:01:41	≡
🔒 Informative Features	Features that pass a "reasonableness" check for useful information, e.g., are not duplicates or reference ids and not a constant value.	30	0	2021-03-31 05:01:41	≡
🔒 Univariate Selections	Features that meet a certain threshold for non-linear correlation with the target (derived after target selection).	24	0	2021-04-10 21:28:55	≡
🔒 DR Reduced Features M8	The most important features based on Feature Impact scores from a particular model.	15	0	2021-04-10 21:36:34	≡

Figure 5.24 – Feature Lists

Here, you will see a list that contains all the raw features, ones that have selections based on univariate analysis (that is, analysis of features one at a time), and also ones that have the most important features. The **DR Reduced Features M8** list or the **Univariate Selections** list look like good starting points. Click on the **Project Data** menu to go back to the data view. Now, let's inspect the univariate list by selecting **Univariate Selections** from the **Feature List** dropdown, as illustrated in the following screenshot:

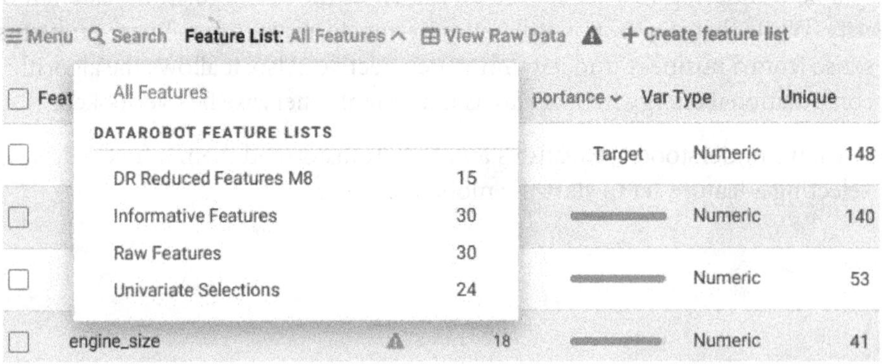

Figure 5.25 – Selecting a feature list

You can now inspect the list of features that have been selected. You can modify this list and create new feature lists by dropping any features that you do not want to include in this list. As you can see, DataRobot has done much of the feature selection for you to get things started. You can remove some more now, or you can remove them in the next iteration after you have built an initial set of models.

Interestingly, DataRobot has already built some models with some of these lists, which we will explore in the next chapter.

Summary

In this chapter, we learned how to bring data into DataRobot. We learned how to assess data quality and to perform EDA by using DataRobot's features. We saw how DataRobot makes it very easy to explore data, set up target features, and perform correlation (or, more accurately, association analysis).

We learned how to leverage DataRobot's output to gain a better understanding of our problem and dataset, and then how to create feature lists to be used in model building. You could do these tasks in Python or R and they are not very difficult, but they do consume some time. This time is better served in focusing on understanding the problem and the dataset.

In the next chapter, we will jump into something that most of you must be waiting for: building models.

6
Model Building with DataRobot

In this chapter, we will see how DataRobot is used to build models. Much of the model-building process has been automated, and DataRobot offers many capabilities to explore a wide range of algorithms automatically, as well as allowing data scientists to fine-tune what they want to build. This results in significant time savings for data science teams and leads to the exploration of many more models than would otherwise be possible. It also leads to better adherence to best practices and hence fewer chances of making mistakes.

By the end of this chapter, you will have learned how to utilize DataRobot to build a wide range of models. In this chapter, we're going to cover the following main topics:

- Configuring a modeling project
- Building models and the model leaderboard
- Understanding model blueprints
- Building ensemble models

Configuring a modeling project

In the previous chapter, we created a project and performed data analysis. We also saw that DataRobot automatically built several models for us. To build these models, we used default project settings.

In this section, we will cover what DataRobot did for us by default and look at how we can fine-tune that behavior. If you remember, once we click the **Start** button on the project page (see *Figure 5.1* in *Chapter 5, Exploratory Data Analysis with DataRobot*), we cannot make any changes to the project options. We will therefore create a new project to review and select the options we want.

For this, let's go into DataRobot and select the **Create New Project** menu option. Just as before, we will now upload the same automobile dataset file that we used before. This time, you can name the project `Automobile Example 2`, as illustrated in the following screenshot:

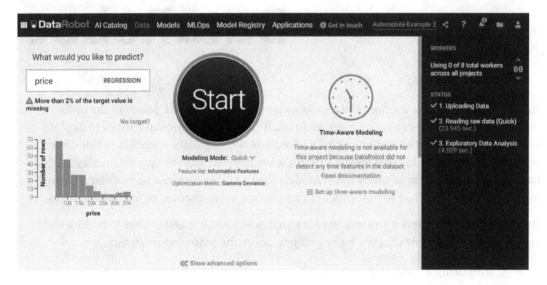

Figure 6.1 – Uploading the dataset for a new project

You can select the same target feature (price) as we did previously. Now, instead of clicking the **Start** button, please click on **Show advanced options** at the bottom of the screen. You will now see the **Advanced Options** screen, as illustrated in the following screenshot:

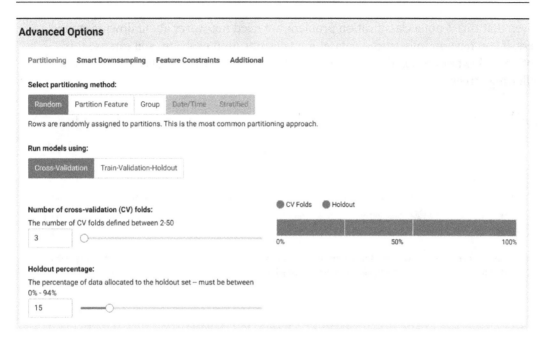

Figure 6.2 – Advanced Options

Here, you can see the partitioning options. You can see the default settings and can change them as needed. Since the amount of data we have is very limited, I have reduced the number of cross-validation folds to 3 and the holdout percentage to 15%. You can easily change these values and run with a different setup as needed. Next, we click on the **Smart Downsampling** tab, as illustrated in the following screenshot:

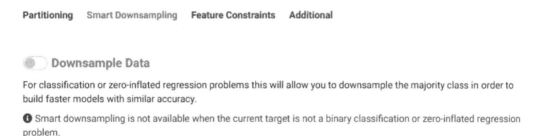

Figure 6.3 – Smart Downsampling

Given that this is not a classification problem, we need not worry about downsampling here. If you have an imbalanced dataset for a classification problem, you can use this option to downsample. Let's now look at the **Feature Constraints** tab, as illustrated in the following screenshot:

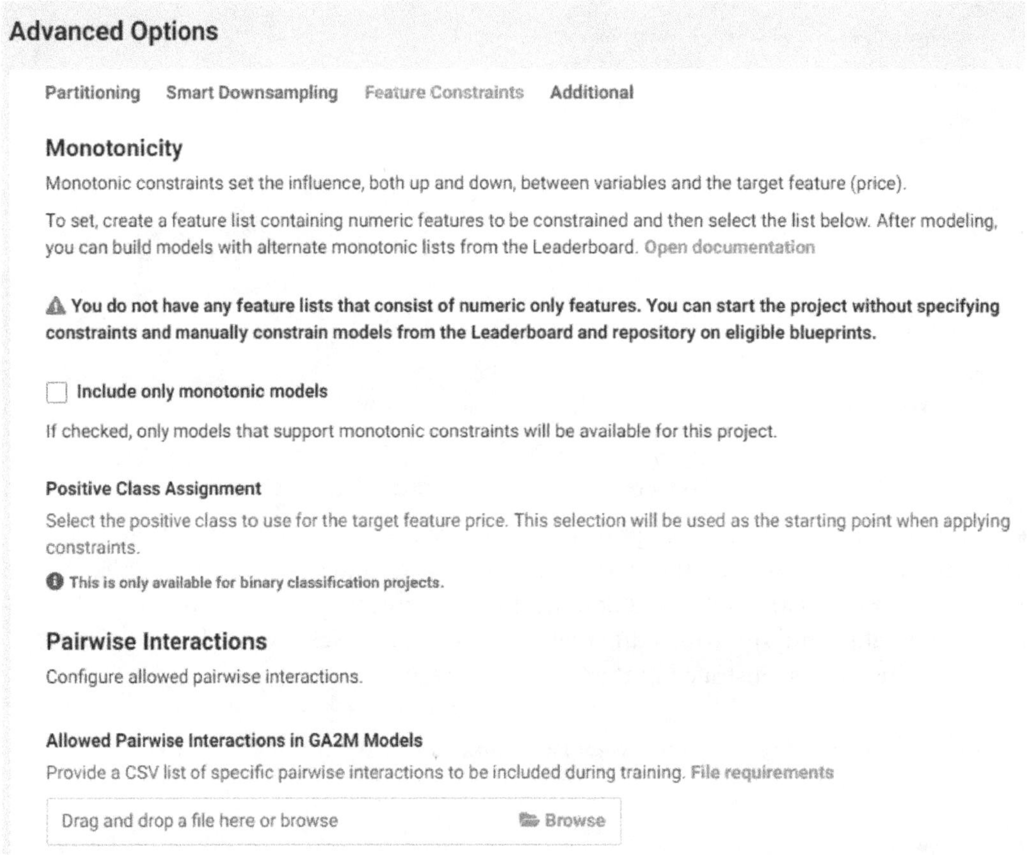

Advanced Options

Partitioning   Smart Downsampling   Feature Constraints   Additional

Monotonicity

Monotonic constraints set the influence, both up and down, between variables and the target feature (price).

To set, create a feature list containing numeric features to be constrained and then select the list below. After modeling, you can build models with alternate monotonic lists from the Leaderboard. Open documentation

⚠ **You do not have any feature lists that consist of numeric only features. You can start the project without specifying constraints and manually constrain models from the Leaderboard and repository on eligible blueprints.**

☐ Include only monotonic models

If checked, only models that support monotonic constraints will be available for this project.

Positive Class Assignment

Select the positive class to use for the target feature price. This selection will be used as the starting point when applying constraints.

ⓘ This is only available for binary classification projects.

Pairwise Interactions

Configure allowed pairwise interactions.

Allowed Pairwise Interactions in GA2M Models

Provide a CSV list of specific pairwise interactions to be included during training. File requirements

    Drag and drop a file here or browse   📁 Browse

Figure 6.4 – Feature Constraints

Here, you can set up constraints on features such as monotonicity—that is, whether the target values move in the same direction as the value of a feature increases. At this point, we do not foresee a need to set such constraints. Such constraints could be part of regulatory requirements in certain use cases. If they are, they can be specified here. Most use cases do not require such a constraint. Let's now click on the **Additional** tab, as illustrated in the following screenshot:

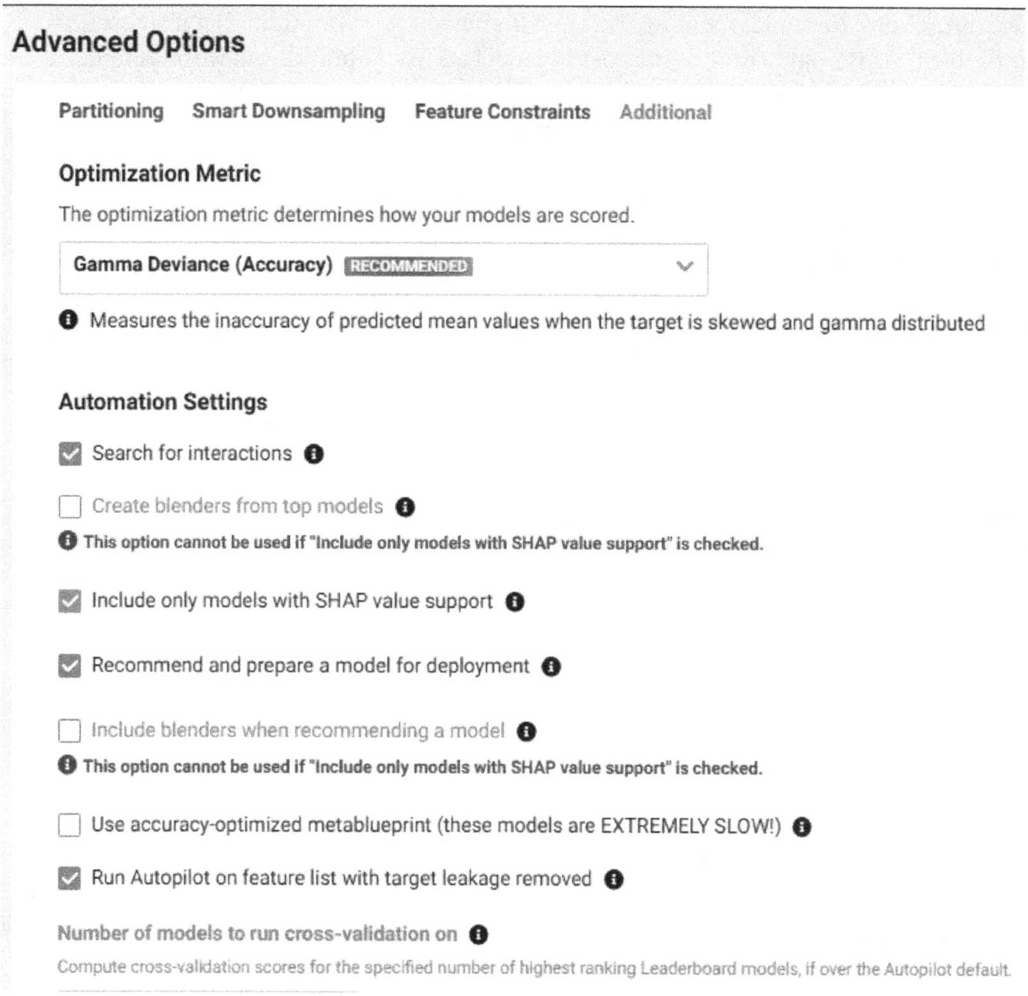

Advanced Options

Partitioning   Smart Downsampling   Feature Constraints   Additional

Optimization Metric

The optimization metric determines how your models are scored.

Gamma Deviance (Accuracy) RECOMMENDED	⌄

ℹ️ Measures the inaccuracy of predicted mean values when the target is skewed and gamma distributed

Automation Settings

☑ Search for interactions ℹ️

☐ Create blenders from top models ℹ️
ℹ️ This option cannot be used if "Include only models with SHAP value support" is checked.

☑ Include only models with SHAP value support ℹ️

☑ Recommend and prepare a model for deployment ℹ️

☐ Include blenders when recommending a model ℹ️
ℹ️ This option cannot be used if "Include only models with SHAP value support" is checked.

☐ Use accuracy-optimized metablueprint (these models are EXTREMELY SLOW!) ℹ️

☑ Run Autopilot on feature list with target leakage removed ℹ️

Number of models to run cross-validation on ℹ️
Compute cross-validation scores for the specified number of highest ranking Leaderboard models, if over the Autopilot default.

Figure 6.5 – Additional options

Here, we see an option to change the **optimization metric** to be used for modeling. I have found DataRobot's recommendations to be very good and you should use this option unless you have a compelling business reason to select a different metric. Given that we are in the early stages of modeling and we are interested in understanding our data, we will select the **Search for interactions** option, unselect the **Create blenders from top models** option, and select the **Include only models with SHAP value support** option. As discussed in *Chapter 2, Machine Learning Basics*, **SHapley Additive exPlanations (SHAP)** values are helpful for understanding the models and will provide additional insights into our problem. This might come at the cost of model accuracy, but we will worry about improving accuracy later. If you scroll down further on this page, you will see even more options, as illustrated in the following screenshot:

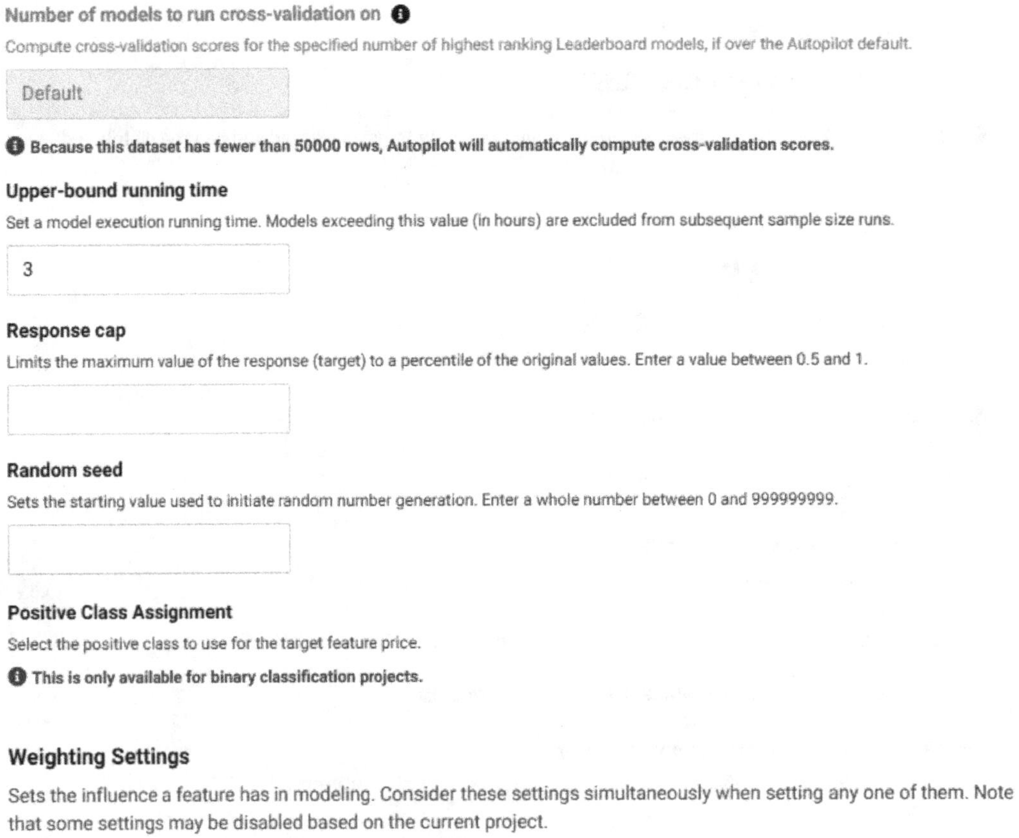

Number of models to run cross-validation on ℹ

Compute cross-validation scores for the specified number of highest ranking Leaderboard models, if over the Autopilot default.

> Default

ℹ **Because this dataset has fewer than 50000 rows, Autopilot will automatically compute cross-validation scores.**

Upper-bound running time

Set a model execution running time. Models exceeding this value (in hours) are excluded from subsequent sample size runs.

> 3

Response cap

Limits the maximum value of the response (target) to a percentile of the original values. Enter a value between 0.5 and 1.

Random seed

Sets the starting value used to initiate random number generation. Enter a whole number between 0 and 999999999.

Positive Class Assignment

Select the positive class to use for the target feature price.

ℹ **This is only available for binary classification projects.**

Weighting Settings

Sets the influence a feature has in modeling. Consider these settings simultaneously when setting any one of them. Note that some settings may be disabled based on the current project.

Figure 6.6 – More options

Here, you can set options to place an upper bound on running time, cap the value of target variable predictions, set a random seed, or add a weighting for a specific feature. For now, we do not see a need to change any of these defaults.

This completes the configuration process, and we are now ready to build the models.

Building models and the model leaderboard

Once we are done making any changes to the configuration settings, we can scroll up and click the **Start** button. DataRobot will now start automatically building the models, as illustrated in the following screenshot:

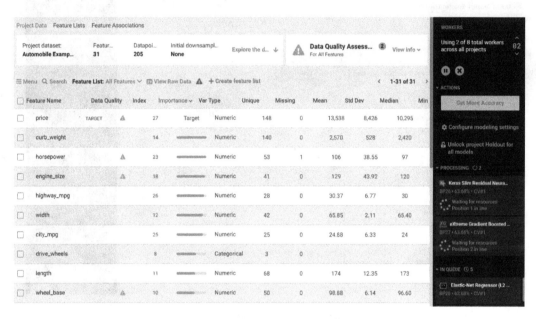

Figure 6.7 – Automated building of models

You can see which models DataRobot is building and how much training data is being used. You will notice that DataRobot will first build quick models with smaller datasets, learn which one performs better, and then selectively build models with more data. In the present case, you might not see this because there is very little data to begin with. Once DataRobot is done building the models, it will show the model leaderboard, as illustrated in the following screenshot:

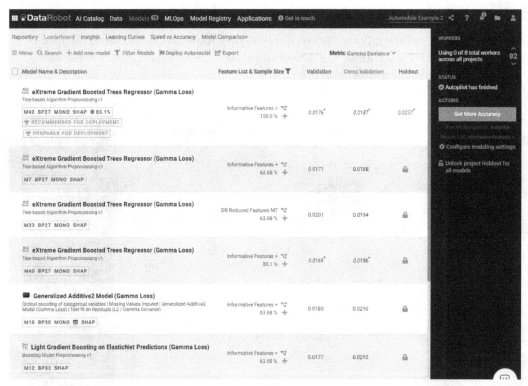

Figure 6.8 – Model leaderboard

In the preceding screenshot, you will see which models rise to the top based on the metric you have selected for cross-validations. You can also choose different metrics from the dropdown to see how the models compare for different metrics. You can clearly see which models rose to the top. It is not uncommon to see gradient boosted models in the top tier. You will also notice that the model rankings change a bit based on the metric selected. You will see that once DataRobot has selected a model for deployment, it has unlocked the holdout results and trained the model with 100% data to prepare for deployment. For now, we will ignore that as we are not yet ready to discuss deployment. Another thing to notice is the feature list used for the top models. You will see that a new **Informative Features** + feature list has been used. This is a feature list that DataRobot created for these models. Let's take a look at what this list contains, as follows:

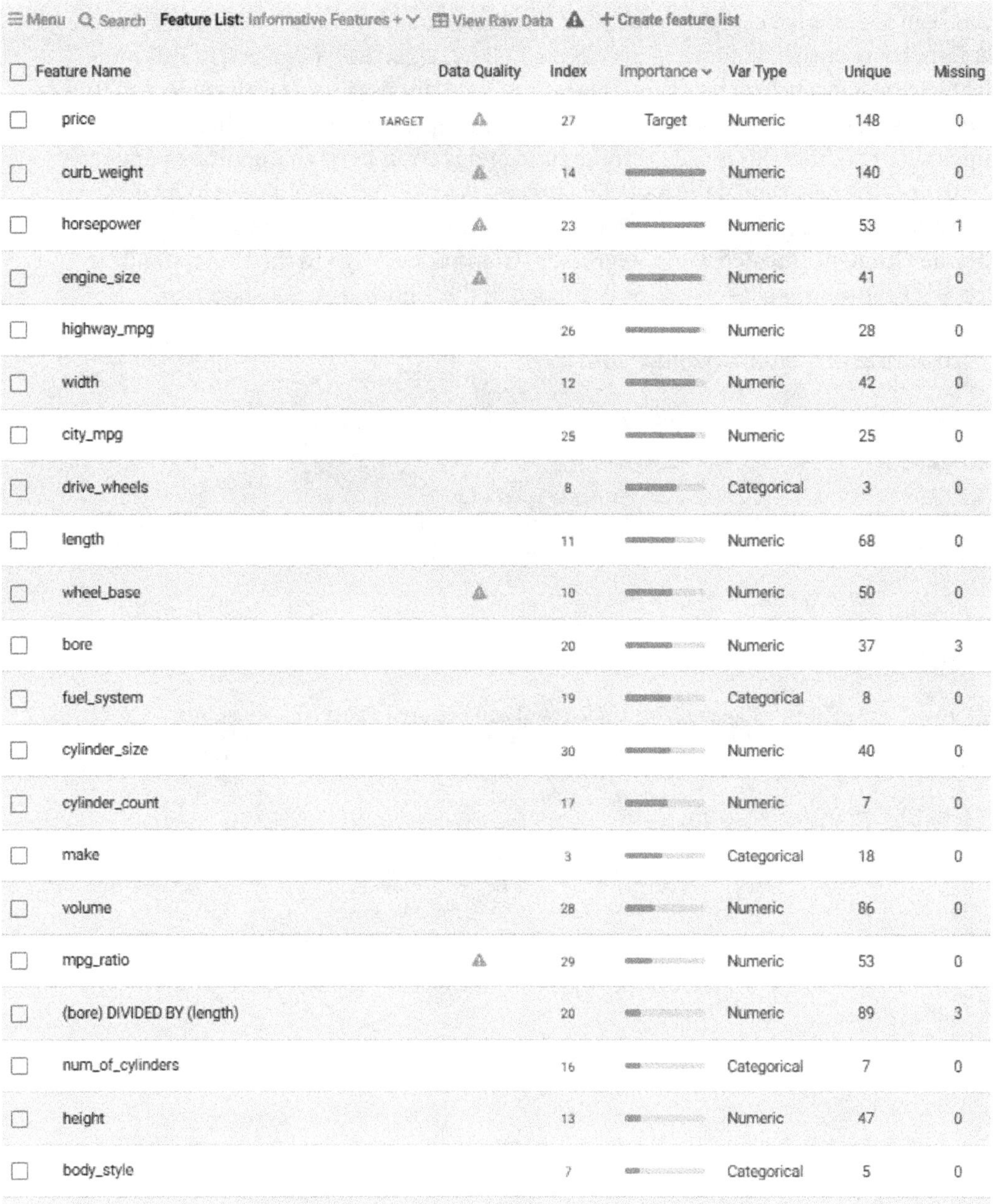

Feature Name	Data Quality	Index	Importance ∨	Var Type	Unique	Missing
price	TARGET ⚠	27	Target	Numeric	148	0
curb_weight	⚠	14	▬▬▬▬▬▬	Numeric	140	0
horsepower	⚠	23	▬▬▬▬▬	Numeric	53	1
engine_size	⚠	18	▬▬▬▬▬	Numeric	41	0
highway_mpg		26	▬▬▬▬▬	Numeric	28	0
width		12	▬▬▬▬	Numeric	42	0
city_mpg		25	▬▬▬▬	Numeric	25	0
drive_wheels		8	▬▬▬	Categorical	3	0
length		11	▬▬▬	Numeric	68	0
wheel_base	⚠	10	▬▬▬	Numeric	50	0
bore		20	▬▬▬	Numeric	37	3
fuel_system		19	▬▬▬	Categorical	8	0
cylinder_size		30	▬▬▬	Numeric	40	0
cylinder_count		17	▬▬	Numeric	7	0
make		3	▬▬	Categorical	18	0
volume		28	▬▬	Numeric	86	0
mpg_ratio	⚠	29	▬▬	Numeric	53	0
(bore) DIVIDED BY (length)		20	▬▬	Numeric	89	3
num_of_cylinders		16	▬▬	Categorical	7	0
height		13	▬	Numeric	47	0
body_style		7	▬▬	Categorical	5	0

Figure 6.9 – New feature list

As you can see, this list contains a subset of the features, and it also contains a new feature that DataRobot created automatically: `(bore) DIVIDED BY (length)`. This ratio might have significance for an engine, and you should discuss its role with **subject-matter experts (SMEs)**. If not previously known, this could represent a new insight for your business team. It turns out that this is called **stroke ratio** and is considered an important parameter for engines. The next step in the modeling process is to see if there is a need to further refine this feature list. Let's go back to the model leaderboard, select the top-performing model **eXtreme Gradient Boosted Trees Regressor (Gamma Loss)**, go to the **Understand** tab and select the **Feature Impact** sub-tab, as illustrated in the following screenshot:

Figure 6.10 – Understand and Feature Impact tabs

You will see that feature impacts are not computed for every model, so go ahead and click on the **Enable Feature Impact** button to let DataRobot compute it. Once clicked, DataRobot will start computing the impacts and show you the results, as illustrated in the following screenshot:

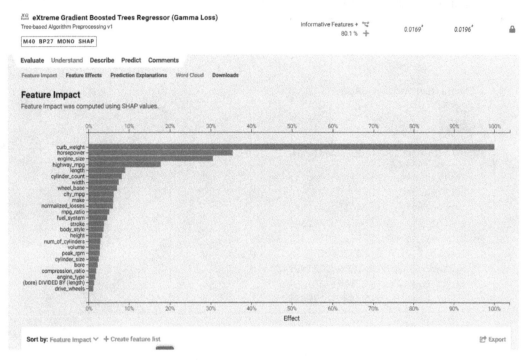

Figure 6.11 – Feature impacts

You will notice that the feature impacts are computed using SHAP values, which we discussed previously. By default, it shows the top 25 features. We will discuss the details of the features and the model later on. For now, we want to look at the entire feature set. For this, we will click on the **Export** button in the bottom-right corner. We will now see the **Export** option, as illustrated in the following screenshot:

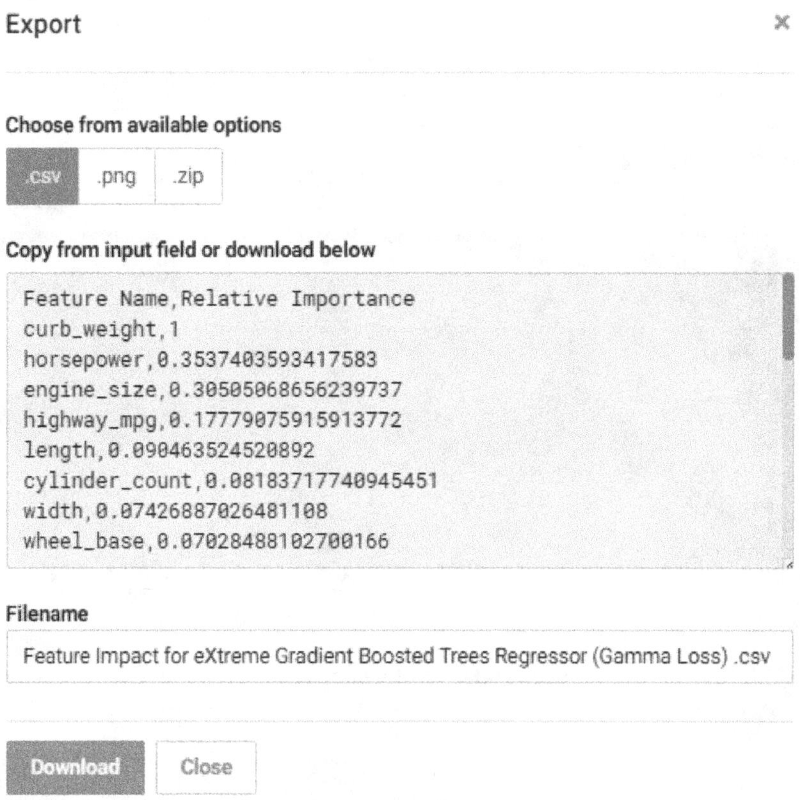

Figure 6.12 – Exporting feature impacts

You can download this information as a `.csv` file to explore it in more detail. Let's use Excel to open the `.csv` file to review the feature impacts, as illustrated in the following screenshot:

Feature Name	Relative Importance
curb_weight	1.00
horsepower	0.35
engine_size	0.31
highway_mpg	0.18
length	0.09
cylinder_count	0.08
width	0.07
wheel_base	0.07
city_mpg	0.06
make	0.06
normalized_losses	0.06
mpg_ratio	0.05
fuel_system	0.05
stroke	0.04
body_style	0.04
height	0.03
num_of_cylinders	0.03
volume	0.03
peak_rpm	0.03
cylinder_size	0.02
bore	0.02
compression_ratio	0.02
engine_type	0.02
(bore) DIVIDED BY (length)	0.01
drive_wheels	0.01
symboling	0.01
aspiration	0.01
num_of_doors	0.00
fuel_type	0.00
engine_location	0.00

Figure 6.13 – Feature impacts of the entire set

As you can see, the last seven features don't add much, and we can try removing them and see the impact. One of the benefits of a tool such as DataRobot is that running these experiments is very quick and easy. Now that we know what we want to do, let's go back to the **Feature Impact** screen. Notice the **+ Create feature list** button on the bottom left. Clicking on that button brings up a dialog box for creating a new feature list, as illustrated in the following screenshot:

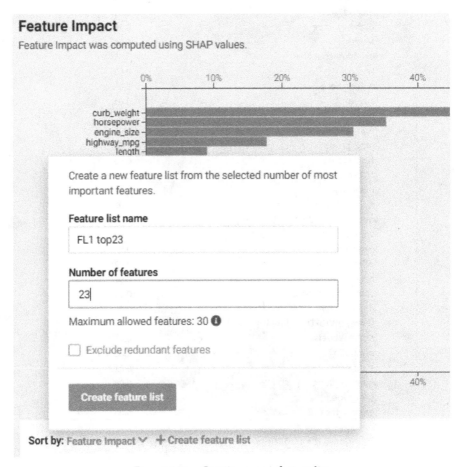

Figure 6.14 – Creating a new feature list

Here, we can give the feature list a new name, FL1 top23, and specify that we want the 23 best features. Now, we can click the **Create feature list** button to save this new feature list. Now that a new feature list has been created, we can now click on **Configure Modeling Settings** in the column on the right side of the page. This will bring up the configuration dialog box, as illustrated in the following screenshot:

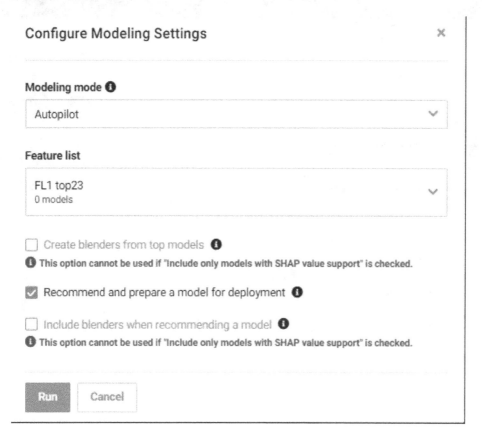

Figure 6.15 – Configure Modeling Settings

We can now select the new feature list, `FL1 top23`, from the dropdown. We can modify the other settings if we need and click the **Run** button. DataRobot will now start building models with the new feature list and when the process completes, you can see the new models in the leaderboard, as illustrated in the following screenshot:

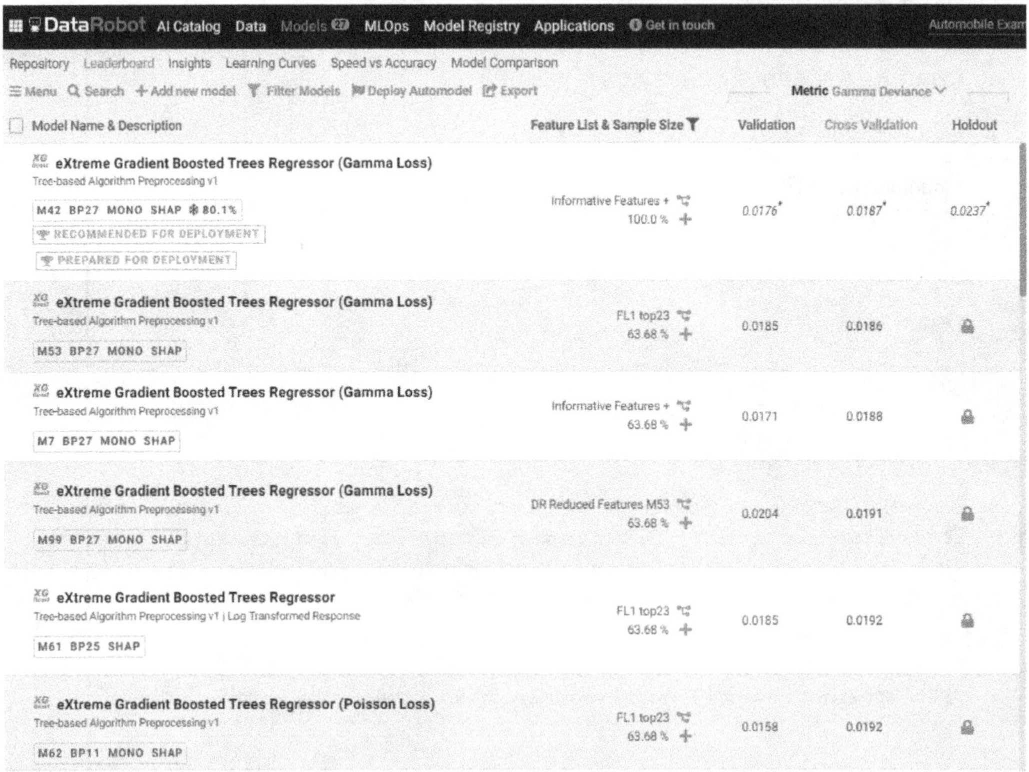

Figure 6.16 – Leaderboard with new models

As you can see, the model built with the new feature list did better and is now at the top of the leaderboard (ignore the deployment-ready model, as it uses the entire dataset). As we can see, removing features that did not contribute much actually helped the model (even if just a little). Given that this model uses a smaller set of features, it is a more desirable model. We can continue this process as needed. At this point, we also start looking more deeply at the model's details and the results it is producing. We will come back to that topic in the next chapter. For now, we want to look at the model blueprints or the steps DataRobot takes to build a model.

Understanding model blueprints

DataRobot performs a lot of data transformations and hyperparameter tuning while building a model. It leverages a lot of best practices to build a specific type of model, and these best practices are codified in the form of blueprints. You can inspect these blueprints to gain insights into these best practices and also to better understand which steps were taken to build a model. To inspect the blueprint for a model, you can click on a model, go to the **Describe** tab, and then select the **Blueprint** tab, as illustrated in the following screenshot:

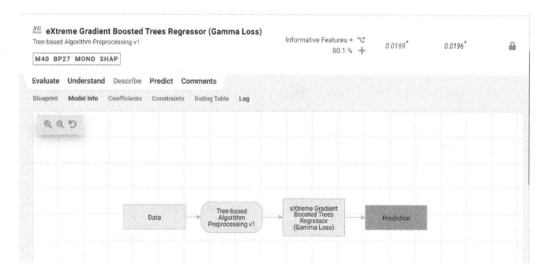

Figure 6.17 – Model blueprint

Here, you can see the workflow steps. As you can see, this blueprint is fairly simple. This is because gradient boost methods are very flexible and do not require a lot of preprocessing. Let's look at another model that did pretty well, the **Generalized Additive2 Model (Gamma Loss)** blueprint, as illustrated in the following screenshot:

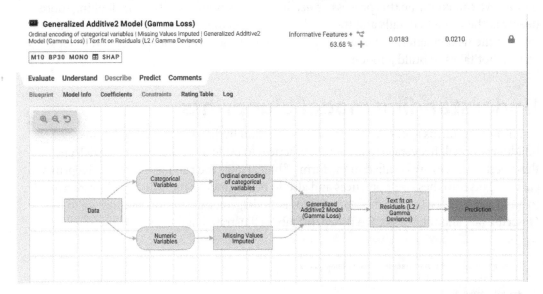

Figure 6.18 – Model blueprint for Generalized Additive2 Model (Gamma Loss)

Here, you can see that preprocessing was required for categorical variables and also for missing values. Let's now look at another blueprint for a **deep learning (DL)** model. Select the **Keras Slim Residual Neural Network Regressor using Training Schedule** model and select the **Blueprint** tab, as illustrated in the following screenshot:

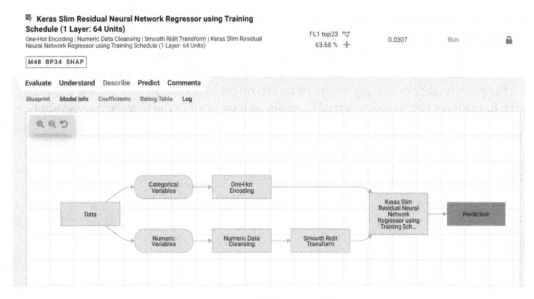

Figure 6.19 – Model blueprint for Keras

You can see that for Keras, we need to perform data cleansing, scaling, and one-hot encoding for categorical variables. You can inspect the details of each of these steps by clicking on the model box, as illustrated in the following screenshot:

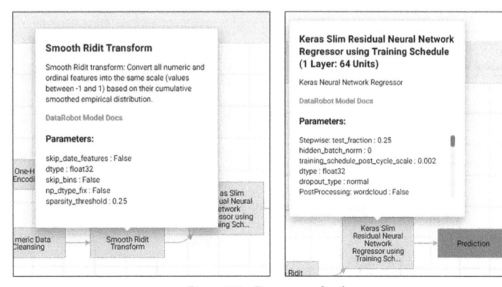

Figure 6.20 – Process step details

You can now see an explanation of which tasks were performed and which hyperparameter settings were used for building the model. There is also a link to additional details about the method used. After inspecting the blueprints, you might see that one of your favorite algorithms was not used by DataRobot, and you might wonder what the performance of that algorithm or model might look like. To do this, you can click on the **Repository** tab at the top left of the page, as illustrated in the following screenshot:

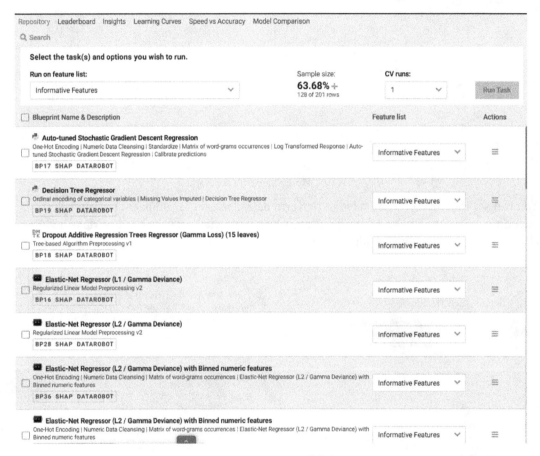

Figure 6.21 – Repository of blueprints

Here, you will see all the blueprints DataRobot has to offer that are relevant to this project. As you can see, this is a pretty comprehensive list. Please note that this list will vary for projects of different types (for example, a time-series project). You can select any one of these blueprints and build a model. The new model will be shown on the leaderboard, where you can assess its relative performance. For now, we are not interested in doing that for this particular project.

We are interested at this point in comparing some of the models to see how well they compare at a more detailed level. For this, let's click on the rightmost tab at the top, called **Model Comparison**. This brings up a page where you can select any two models to see how they match up, as illustrated in the following screenshot:

Figure 6.22 – Model Comparison

Here, we have selected the XGBoost model and the **generalized additive model** (**GAM**) model for comparison across multiple metrics. We can see that the two models are not too far apart, and you can select either one depending on other factors. As we discussed previously, GAMs have the advantage of being easy to explain to business users and can be presented as a factor **lookup table** (**LUT**), sometimes called a rating table. There might also be regulatory reasons to select a GAM model. Let's explore a bit further by clicking on the **Compute dual lift data** button, to take us to the following screen:

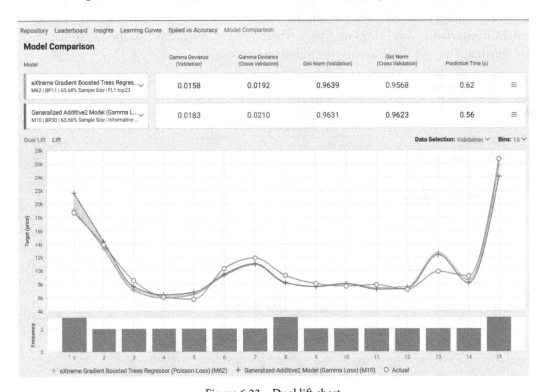

Figure 6.23 – Dual lift chart

Dual lift charts are used to compare the results of two models. For a dual lift chart, the results are sorted by the difference between the two models as opposed to the target value. The values are then binned to display the results for each bin. The shaded area depicts the difference between the two models. Here, again, we see that the two models are very similar in their performance.

If two models have overall good scores but show large deviations in values in this chart, then these models will be good candidates for creating an ensemble model.

Building ensemble models

It is well known that ensembles of models tend to perform better and also tend to be more robust. DataRobot provides the capability to automatically build ensemble models; however, this does require some trade-offs. For example, ensemble models take more time and computational resources to build and deploy, and they also tend to be more opaque. This is the reason we did not start off by building ensemble models. Once you have built several models and you are interested in ways of improving your model accuracy, you can decide to build ensembles. As we saw in the previous sections, we have to explicitly select the option to build ensembles, and that also means that we cannot compute SHAP values. Let's look at how this is done. Let's first go to the project list page, which shows all of your current projects, as illustrated in the following screenshot:

Figure 6.24 – Project list

Here, we will select the **Actions** icon for the project that we have been working on, which is `Automobile Example 2`. From the menu, we will select the **Duplicate** option. You will now see the **Duplicate** dialog box, as illustrated in the following screenshot:

Figure 6.25 – Duplicating a project

We can give it a new name, `Automobile Example 3`, and we will select **Copy dataset only**. This way, we can apply new project settings. Let's click **Confirm**. This will create a new project. We can select the target as price, and now we click on the **Advanced Options** tab, as illustrated in the following screenshot:

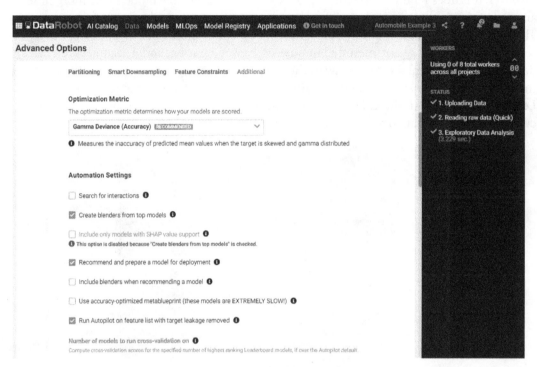

Figure 6.26 – Advanced options for ensembles (blenders)

This time, we will select the **Create blenders from top models** option and uncheck the **Include only models with SHAP value support** option. Now, we can click the **Start** button to let DataRobot build the models. Once DataRobot has finished building the models, we can inspect the leaderboard, as illustrated in the following screenshot:

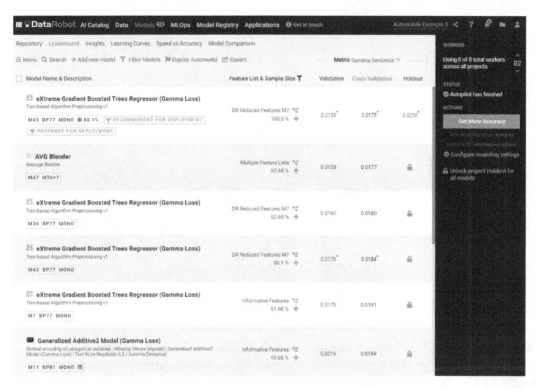

Figure 6.27 – Leaderboard with ensemble models

You will notice the DataRobot has built an **AVG Blender** model that seems to be the top model, but not by much. Blended models can sometimes produce substantial lift over individual models, so it is worthwhile exploring this option. We can select this model and click on the **Describe** tab and then the **Blueprint** tab, as illustrated in the following screenshot:

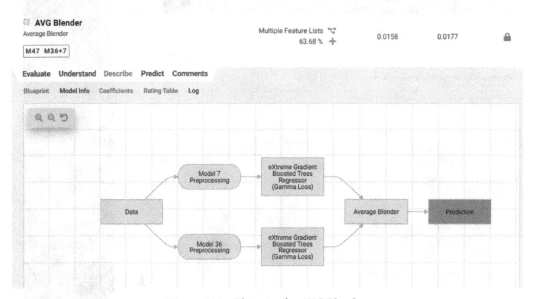

Figure 6.28 – Blueprint for AVG Blender

We can now see that the blender has selected two XGBoost models, and hence it is not surprising that the lift is not much better. In this case, we will not select the blended model, and we go back to the previous project.

Summary

In this chapter, we learned how to build and compare models by leveraging DataRobot's capabilities. As you saw, DataRobot makes it very easy to build many models quickly and helps us compare those models. As you experienced, we tried many things and built dozens of models. This rapid model exploration is DataRobot's key capability, and its importance to a data science team cannot be overstated. If you were to build these models on your own in Python, it would have taken a lot more time and effort. Instead, we used that time and thinking to experiment with different ideas and put more energy toward understanding the problem. We also learned about blueprints that encode best practices. These blueprints can be useful learning tools for new and experienced data scientists alike. We also learned how DataRobot can build ensemble or blended models for us.

It might be tempting to jump ahead and start deploying one of these models, but it is important to not directly jump into that without doing some analysis. We are now ready to dig deeper into the models, understand them, and see if we can gain more insights from them.

7
Model Understanding and Explainability

In the last chapter, we learned how to build models, and we will now learn how to use output generated by DataRobot to understand the models and also use this information to explain why a model provides a particular prediction. As we have discussed before, this aspect is critically important to ensure that we are using the results correctly. DataRobot automates much of the task of creating charts and plots to help someone understand a model, but you still need to know how to interpret what it is showing in the context of the problem you are trying to solve. This is another reason why we will need people involved in the process, even if much of a task has been automated. As you can imagine, the task of interpreting the results will therefore become more and more valuable as the degree of automation increases.

In this chapter, we're going to cover the following main topics:

- Reviewing and understanding model details
- Assessing model performance and metrics
- Generating model explanations
- Understanding model learning curves and trade-offs

Reviewing and understanding model details

In the last chapter, we created several models for different projects. DataRobot creates 10 to 20 models in a project, and it would be very onerous to look at and analyze the details of all of these models. You do not have to review each of these models, and it is common to review only the top few models before making a final selection. We will now look at the leaderboard for models in the `Automobile Example 2` project and select the top model, as illustrated in the following screenshot:

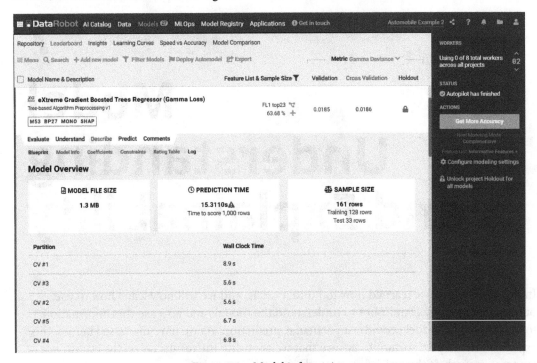

Figure 7.1 – Model information

In the preceding screenshot, we selected the **Model Info** tab within the **Describe** tab to get a view of how large the model is and the expected time it takes to create predictions. This information is useful in real-time applications that are time-sensitive and need to score thousands of transactions quickly. Let's now go to the **Feature Impact** tab within the **Understand** tab, as shown in the following screenshot:

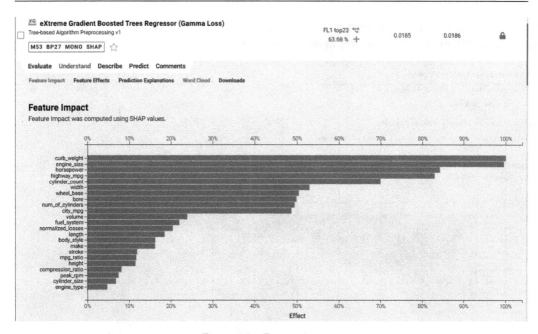

Figure 7.2 – Feature impacts

This is one of the most important charts for the model as it shows how much a feature contributes to this XGBoost model. We can see that the top contributors are `curb_weight`, `engine_size`, `horsepower`, `highway_mpg`, and `cylinder_count`. On the other hand, `cylinder_size` and `engine_type` contribute very little. While it is true that `cylinder_size` is not very predictive, we must not forget that prediction is not always the end objective. We know that `cylinder_size` has an effect on `engine_size`, an important feature. The objective might be to use this information to figure out ways to reduce costs. For that, we might want to reduce `engine_size`, but you cannot reduce `engine_size` directly. For that, you need to reduce the size or count of cylinders, which will lead to a reduction in `engine_size`. Having a causal diagram of this problem to guide you becomes very helpful in determining the best actions to take to achieve our objectives.

Before we take action, let's inspect what the results look like for a **Generalized Additive Model (GAM)**, as shown in the following screenshot:

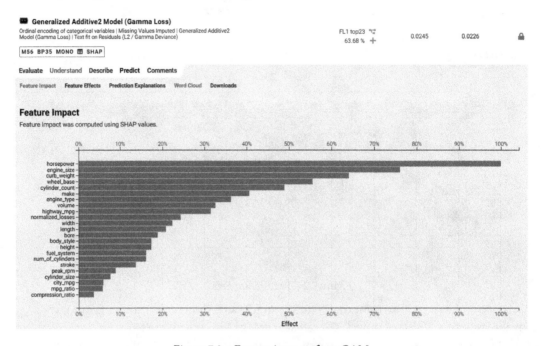

Figure 7.3 – Feature impacts for a GAM

Figure 7.3 shows the important features of the GAM. While many of the features look similar, we notice that `engine_type` is fairly high in importance for this model, whereas `engine_type` was very low in importance for the previous model. This is not an error— it points to the fact that many of the features are interrelated and different models can pick up signals from different features, and that predictive power is not necessarily the same as the root cause. To take action, we need to understand the root feature that leads to a change in the target feature. To put this another way, the feature that best predicts something is not always the feature that can be changed to create the desired change in the target.

To further understand how a feature affects the target, let's select the **Feature Effects** tab within the **Understand** tab of the model, as shown in the following screenshot:

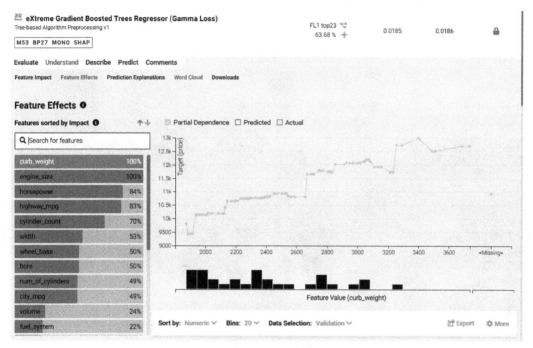

Figure 7.4 – Feature effects

The preceding screenshot shows partial dependence plots for various features. The selected plot is for `curb_weight`. The plot shows a fairly linear relationship between `curb_weight` and price. We do see some unusual dips in price in a few spots—for example, around a `curb_weight` value of `2700`. Before we take that too seriously, we notice that the amount of data around that is very limited. This tells us that this particular observation is likely due to a lack of data. This does raise the issue that our model is likely to predict a lower price in that small region, which in turn could result in lower revenue.

Let's look at another feature in the following screenshot:

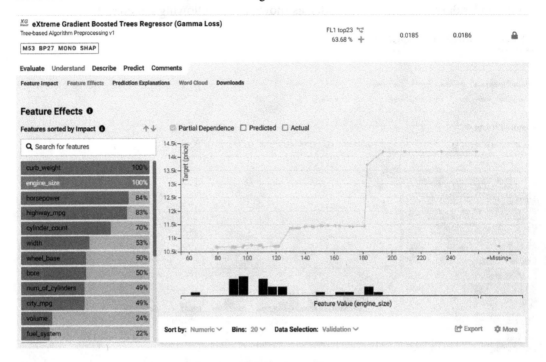

Figure 7.5 – Partial dependence for engine_size

The preceding screenshot shows a highly non-linear relationship between `engine_size` and price. We see a very dramatic rise in price around the `engine_size` value of `180`. It is hard to know how real this effect is without discussing it with domain experts. We can notice that the amount of data available for sizes greater than `130` is very small, hence the effects we see could be simply due to a lack of data. Taken as is, it indicates that prices stagnate beyond a size of `200`, and this could be an important insight for the business.

Let's take a look at another partial dependence plot for `highway_mpg` in the following screenshot:

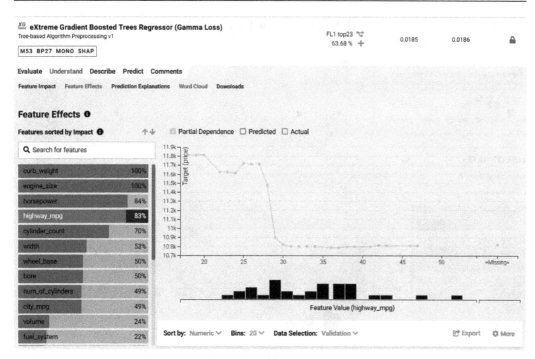

Figure 7.6 – Partial dependence plot for highway_mpg

Figure 7.6 shows another highly non-linear relationship, with a key transition point happening around a `highway_mpg` value of 28. This clearly shows a big price drop around 28, hence this is a critical point. This could be due to regulations, where going below 28 places you in a different type of vehicle or engine. We also notice that once you get above that threshold, any further change is not very meaningful from a price impact (however, it could still be very impactful from other perspectives). If you do not know why this is, it is important for you to discuss this with your **subject-matter experts (SMEs)**.

My main objective for showing and discussing these plots is to show you how important it is to spend your time analyzing and reviewing these plots rather than spending all of your time coding up these plots. Since DataRobot automatically creates these for you, you can now spend your time doing the more value-added work of analyzing these results to help improve your business.

Let's revisit the `engine_size` plot, but this time for the GAM, as shown in the following screenshot:

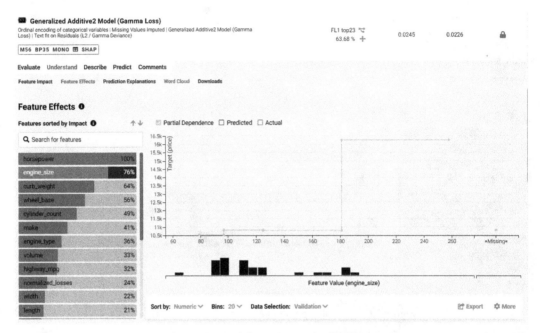

Figure 7.7 – Partial dependence plot for the GAM

Figure 7.7 shows the partial dependence for the GAM. Comparing this with *Figure 7.5*, we see that *Figure 7.7* shows clearer thresholds around values `95` and `180`. Discussing this with domain experts could help you determine which model is a better representation of reality and which model helps you to better set pricing. One of the benefits of GAMs is that you can easily smooth out these curves and shape them for deployment. Remember— accurate prediction is not always the same as better intervention or action.

GAMs are a lot easier to understand and explain. Let's look at another chart here that helps in that understanding:

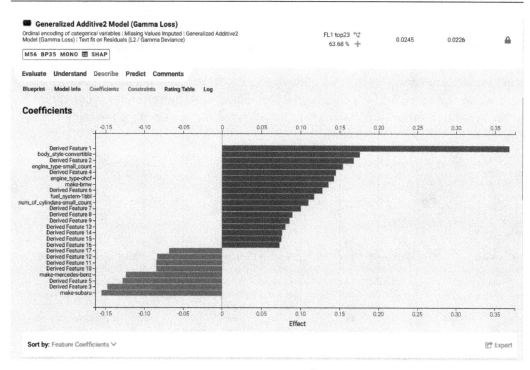

Figure 7.8 – Feature coefficients

Figure 7.8 shows the coefficients for different features in the GAM. You will notice that DataRobot has created some derived features. You can click on them to see more details. This provides a high-level view of the coefficients, but there is another view that provides a better view for understanding the model. For that, let's click on the **Rating Table** tab within the **Describe** tab for the GAM, as shown in the following screenshot:

Figure 7.9 – Rating table for a GAM

This view lets you download the rating table built by DataRobot; you can also modify this table and upload it back to use the modified table. This mechanism thus allows you to manually fine-tune your model based on your understanding of the problem. This feature is therefore very powerful as it allows you a lot of flexibility, but at the same time, you must use this carefully. Let's click on the **Download table** button and download the **comma-separated values (CSV)** file. Once downloaded, we can open the file using **Excel**, as shown in the following screenshot:

Total weights: 128
Intercept: 9.40625052147
Base: 12164.1760469
Model precision: single
Loss distribution: Gamma Deviance
Link function: log

Feature Name	Feature Strength	Type	Transform1	Value1	Weight	Coefficient	Relativity
body_style	0.025250775	CAT	One-hot	'convertible'	5	0.176039263	1.192484879
body_style	0.025250775	CAT	One-hot	'hardtop'	5	0.058601148	1.060352233
body_style	0.025250775	CAT	One-hot	'hatchback'	42	-0.008984998	0.991055247
body_style	0.025250775	CAT	One-hot	'sedan'	62	-0.004507467	0.995502676
body_style	0.025250775	CAT	One-hot	'wagon'	14	-0.036883514	0.963788397
body_style	0.025250775	CAT	One-hot	Other categories	0	0	1
bore	0.027491111	NUM	Binning	(-inf, 3.48]	80	-0.015496588	0.984622866
bore	0.027491111	NUM	Binning	(3.48, inf)	45	0.028582595	1.028994997
bore	0.027491111	NUM	Binning	Missing Value	3	-0.00703511	0.992989579
city_mpg	0.008655331	NUM	Binning	(-inf, 17]	19	0.01255012	1.012629203
city_mpg	0.008655331	NUM	Binning	(17, 23]	36	0.004546997	1.00455735
city_mpg	0.008655331	NUM	Binning	(23, inf)	73	-0.005508824	0.994506321
city_mpg	0.008655331	NUM	Binning	Missing Value	0	-0.005508824	0.994506321
compression_ratio	0.005233876	NUM	Binning	(-inf, 16.0]	116	0.002096349	1.002098548
compression_ratio	0.005233876	NUM	Binning	(16.0, inf)	12	-0.020264704	0.979939246
compression_ratio	0.005233876	NUM	Binning	Missing Value	0	0.002096349	1.002098548
curb_weight	0.093592487	NUM	Binning	(-inf, 1977]	12	-0.127875885	0.879962589
curb_weight	0.093592487	NUM	Binning	(1977, 2133]	14	-0.083281719	0.920091903
curb_weight	0.093592487	NUM	Binning	(2133, 2291]	15	-0.053519981	0.947887001
curb_weight	0.093592487	NUM	Binning	(2291, 2389]	14	-0.045751355	0.955279458
curb_weight	0.093592487	NUM	Binning	(2389, 2665]	20	-0.010236136	0.989816075

Figure 7.10 – Rating table for a GAM

You can now see what the rating table looks like. Here, you see that DataRobot has created bins for various features. For each bin, it has assigned the coefficient and relativity as to how changes in a feature impact the target variable. To understand this a bit better, we can create plots for individual features in Excel, as shown in the following screenshot:

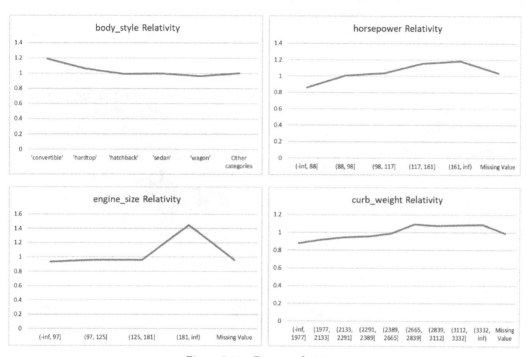

Figure 7.11 – Feature relativities

In *Figure 7.11*, you can see how a given feature such as `body_style` contributes to the price. The GAM model is essentially a sum of all the contributions from the selected features. Given the rating table, anyone can easily calculate the price, and this can also be implemented in a very simple manner. Given that the individual feature effects are non-linear (and still very understandable), this allows these models to perform very well while still being very easy to understand. It is no wonder that GAMs are becoming very popular.

There is one more chart that we want to look at that is frequently helpful in understanding the contributions of features. For this, we will click on the **Insights** menu item at the top of the page, which brings up the chart shown here:

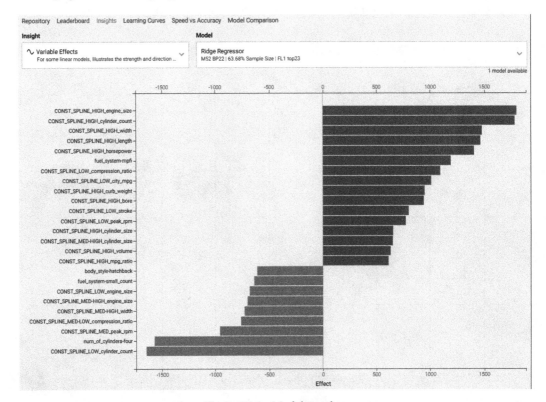

Figure 7.12 – Model insights

Figure 7.12 shows the variable effects using a DataRobot selected model that is built using constant splines (in this case, the **Ridge Regressor** model). This shows the effects of the key feature values in one view, and you can get a sense of relative impact as well as the positive versus the negative contribution of features. A **constant spline** is a feature transformation where a numeric feature is converted into pieces made up of constant splines. The value of the feature is one if the value falls within a specific interval; otherwise, it is zero. You can review this chart with reference to the feature effects for the models you have selected to see if there are any inconsistencies between these charts.

Now that we understand the model from the perspective of which features are important and how they contribute toward the target value, we can focus on how well the model is doing.

Assessing model performance and metrics

In this section, we will focus on how well a model is doing in trying to predict the target values. Let's start by looking at the overall performance comparison across different models, as shown in the following screenshot:

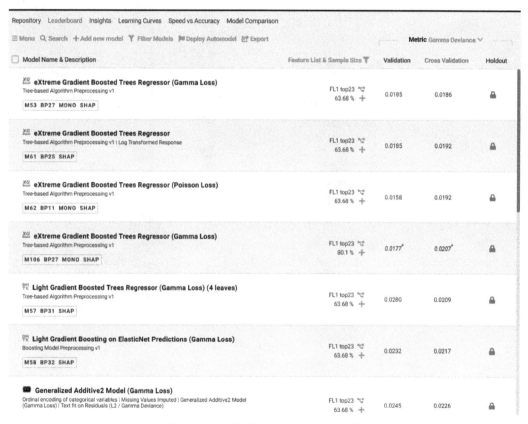

Figure 7.13 – Performance across models

The preceding screenshot shows the overall leaderboard, which we have seen before. Here, we can see the overall performance of different models based on the **Gamma Deviance** metric. We can also review the performance based on other metrics by clicking on the drop-down arrow near the metric, which shows us a list of metrics we can choose from, as illustrated in the following screenshot:

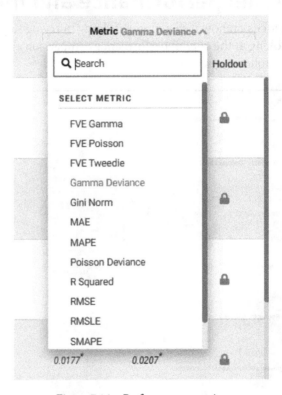

Figure 7.14 – Performance metrics

Figure 7.14 shows the various metrics we can select from. You will typically see a similar trend across different metrics in terms of which models surface to the top spots. In general, the metric that DataRobot selects is a very good choice, if not the best choice. Let's now inspect the performance details of specific models by clicking on the model and selecting the **Lift Chart** tab within the **Evaluate** tab, as shown in the following screenshot:

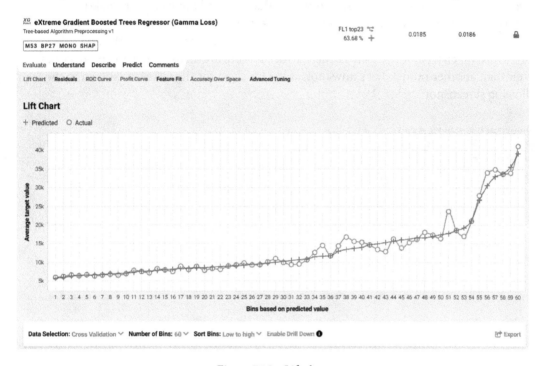

Figure 7.15 – Lift chart

The lift chart illustrated in the preceding screenshot shows how the predictions stack up against the actual values. You can select the number of bins to aggregate the results. The maximum value is 60, and that is normally a good starting point. This means that the predictions are first sorted in ascending order and then grouped into 60 bins. The results you see are the average values within that bin. The reason for binning is that if you look at the entire dataset, there will be so much data that you will not be able to make any sense out of it. You can see that the model does very well over the entire range of values, with some small pockets where the differences seem higher than the rest. We typically want to see lift charts for multiple models, to see if there are areas where one model does better than another model. Let's now look at the lift chart for the GAM, as shown in the following screenshot:

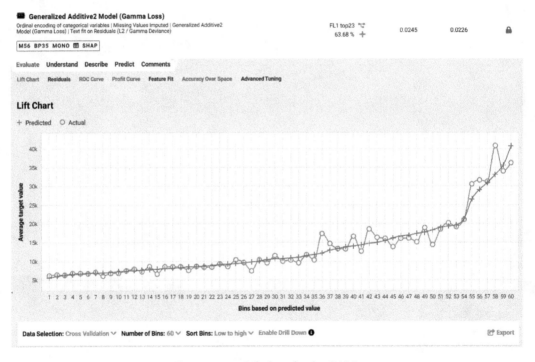

Figure 7.16 – Lift chart for the GAM

The results in *Figure 7.16* look very similar to the results from *Figure 7.15*, but we can see that the GAM did not do as well for higher values. We now know where specifically the GAM is weaker as compared to the XGBoost model. Let's look further by clicking on the **Residuals** tab within the **Evaluate** tab, as shown in the following screenshot:

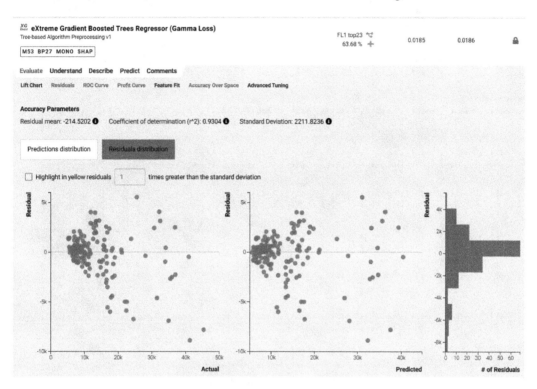

Figure 7.17 – Model residuals

The residuals seem to be well distributed around the mean but with a small skew toward
-ve values. Let's also check how the residuals are distributed for the GAM. We can see the
output in the following screenshot:

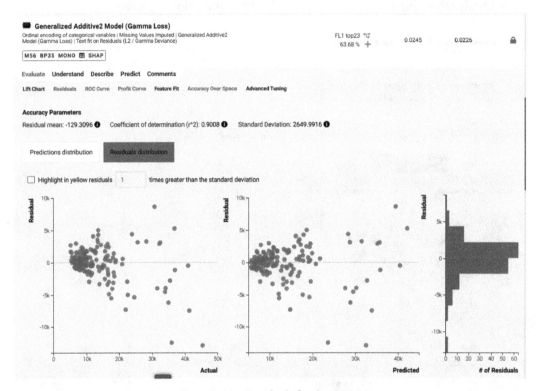

Figure 7.18 – Residuals for the GAM

The residuals for the GAM are also well distributed but with a slightly larger skew
compared to the XGBoost model. Overall, the performance of the models looks very
good. We can now look into understanding individual predictions and their explanations.

Generating model explanations

Another key capability of DataRobot is that it automatically generates instance-level explanations for each prediction. This is important in understanding why a particular prediction turned out the way it did. This is not only important for understanding the model; many times, this is needed for compliance purposes as well. I am sure you have seen explanations generated or offered if you are denied credit. The ability to generate these explanations is not straightforward and can be very time-consuming. Let's first look at the explanations generated for the XGBoost model, as shown in the following screenshot:

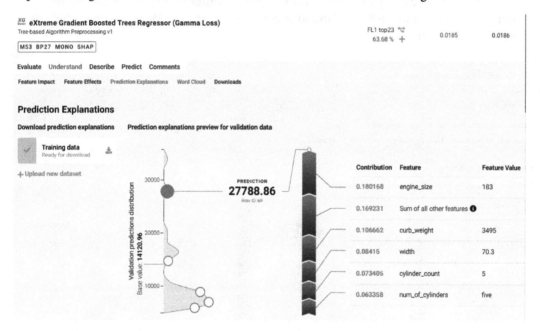

Figure 7.19 – Model explanations

Since we selected the **SHAP** option for this project, the model explanations are based on **SHapley Additive exPlanations (SHAP)** algorithms. Here, you can see the overall distribution of predictions on the left, and you can see that most of the dataset lies in the range of 0 to 10000. You can select some specific points and see the components that make up that prediction. In *Figure 7.19*, we have selected the prediction point of 27788.86. We can see the top contributing elements on the right, where engine_size is contributing the most, and in this case, the value of engine_size is 183. Notice that the relative contribution of features can vary on a case-by-case basis, and the ordering of features here will not exactly match the feature-impacts order we saw in the preceding section. Let's compare this with explanations generated by the GAM, as shown in the following screenshot:

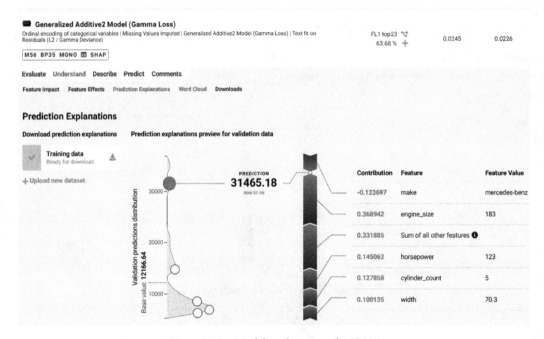

Figure 7.20 – Model explanations for GAM

In the preceding screenshot, the point selected is for a prediction of `31465.18`. For this point, we can see the features that are the main contributors toward that price, and we also note that there was a reduction or -ve contribution due to the **make** of the vehicle being **Mercedes-Benz**. We can also see that in this case, the contribution of `engine_size` of `183` is much larger for the GAM.

The explanations for the entire dataset can be downloaded and analyzed for additional insights. You can also upload an entirely new dataset to score it and generate these explanations very easily, by clicking on the **Upload new dataset** button.

As you have seen in this chapter, different models have different performance, use the features a little bit differently, and have different levels of understandability. There are a few other dimensions that should be looked at before making a final selection of the model you want to use. Let's now look at model learning curves and some of the model trade-offs.

Understanding model learning curves and trade-offs

In **machine learning** (ML) problems, we are always trying to find more data to improve our models, but as you can imagine, there comes a time when we reach a point of diminishing returns. It is very hard to know when you have reached that point, but you can get indications by looking at the learning curves. Fortunately, DataRobot makes that task easy by automatically building these learning curves. When DataRobot starts building models, it first tries a broad range of algorithms on small samples of data. Promising models are then built with bigger sample sizes, and so on.

In this process, we discover how much performance improvement happens as more data is added. To look at the learning curves, you can click on the **Learning Curves** menu item at the top of the screen, as seen in the following screenshot:

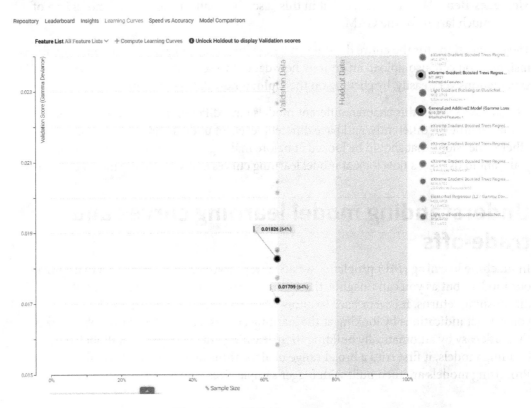

Figure 7.21 – Model learning curves

You can see the different model types on the right-hand side of the page. Here, you can click on the models you want to inspect and compare. After selecting the models, you click on the + **Compute Learning Curves** button. This brings up a dialog box showing the selected models and corresponding sample sizes, as shown in the following screenshot:

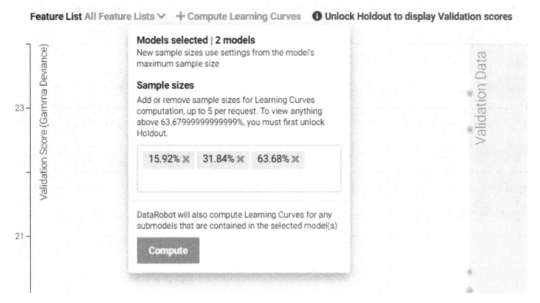

Figure 7.22 – Models selected for comparison

If the selections in *Figure 7.22* look correct, you can click the **Compute** button. You will now see the learning curves for the selected models, as shown in the following screenshot:

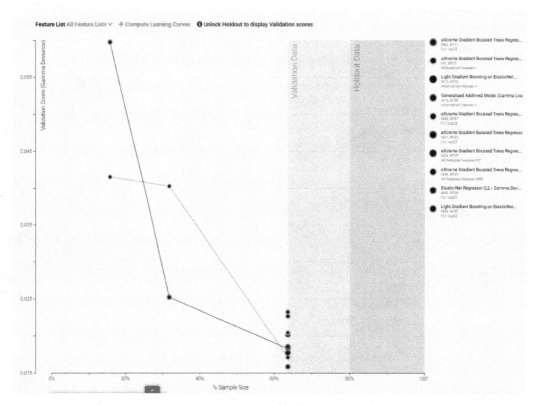

Figure 7.23 – Comparison of learning curves

You can now see the improvement in performance as the sample size increases. We can see that the GAM learns very rapidly, but as the sample size increases, the XGBoost model takes over. We can see that both models will benefit from additional data. We can also see that if we only had half of the data we currently have, then the GAM would have been the clear winner.

We can now look at another trade-off for models—namely, the trade-off between speed and accuracy. If you click on the **Speed vs Accuracy** menu item at the top of the page, you will see a chart, as shown in the following screenshot:

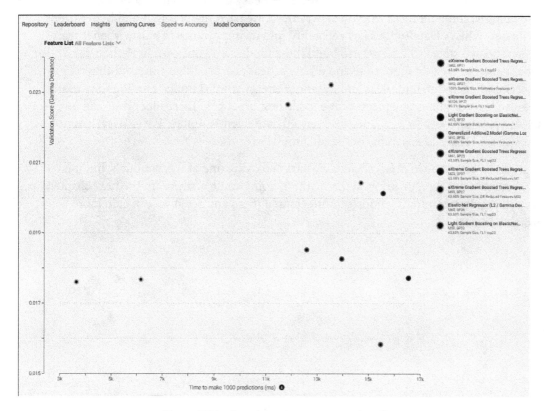

Figure 7.24 – Speed versus accuracy trade-off

You will notice the DataRobot has built an AVG Blender model that seems to be the top model, but not by much. Blended models can sometimes produce substantial lift over individual models, so it is worthwhile exploring this option. We can select this model and click on the **Blueprint** tab within the **Describe** menu item.

Summary

In this chapter, we covered how to build and compare models by leveraging DataRobot's capabilities. As you saw, DataRobot makes it very easy to build many models quickly and helps us compare them. As you experienced, we tried many things and built dozens of models. This is DataRobot's key capability, and its importance to a data science team cannot be overstated. If you were to build these models on your own in Python, it would have taken a lot more time and effort. Instead, we used that time and thinking to experiment with different ideas and put more energy toward understanding the problem. We also learned about blueprints that encode best practices. These blueprints can be useful learning tools for new and experienced data scientists alike. We also learned how DataRobot can build ensemble or blended models for us.

It might be tempting to jump ahead and start deploying one of these models, but it is important to not directly jump to that without doing some analysis. In the next chapter, we will dig deeper into the models to understand them and see if we can gain more insights from them.

8

Model Scoring and Deployment

In the previous chapter, we learned how to use outputs generated by DataRobot to understand models and why a model provides a particular prediction. We will now learn how to use models to score input datasets and create predictions to be used in the intended applications. DataRobot automates many tasks that are required for scoring and generating row-level explanations.

Creating predictions, however, is not where these tasks end. In most cases, these predictions need to be transformed into actions for consumption by people or applications. This mapping of predictions to actions requires an understanding of business and therefore needs a person to interpret the results (in most use cases). In this chapter, we will discuss how this is done. We're going to cover the following main topics:

- Scoring and prediction methods
- Generating prediction explanations
- Analyzing predictions and postprocessing
- Deploying DataRobot models
- Monitoring deployed models

Scoring and prediction methods

DataRobot provides multiple methods to score datasets using models that have been created. One of the easiest methods is batch scoring via the DataRobot **user interface (UI)**. For this, we need to follow these steps:

1. Create a file with the dataset to be scored. Given that we are using a public dataset, we will simply use the same dataset to score. In a real project, you will have access to a new dataset for which you want to create predictions. For our purposes, we simply created a copy of our `imports-85-data.xlsx` dataset file and named it `imports-85-data-score.xlsx`.

2. Now, let's select the **Predict** tab and then the **Test Predictions** tab for the **XGBoost (XGB)** models, as shown in the following screenshot:

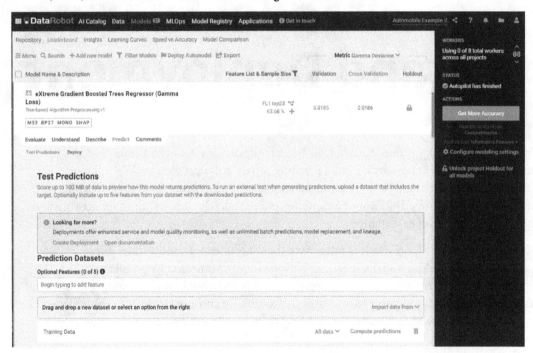

Figure 8.1 – Batch scoring

In the preceding screenshot, you will see that you have an option to **drag and drop a new dataset** to add the scoring file to the model.

3. Let's select our `imports-85-data-score.xlsx` scoring file and drop it into the **Drag and drop a new dataset** box. Once you drop the file, it will get uploaded and you can see it in the interface, as shown in the following screenshot:

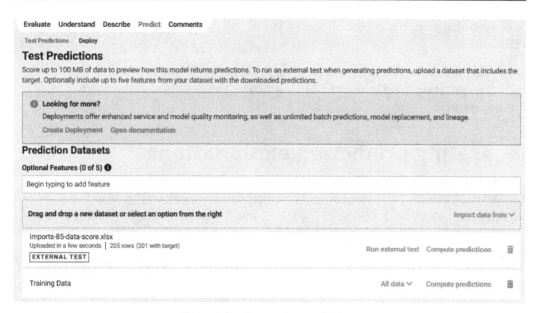

Figure 8.2 – Computing predictions

4. You can now click on the **Compute predictions** button to start the scoring process. Once this process is complete, you can click on the **Download predictions** button to download the predictions generated by the model. The download is in the form of a `.csv` file that you can view in Excel, as shown in the following screenshot:

	A	B	C
1	row_id	Prediction	
2	0	15158.04199	
3	1	15158.04199	
4	2	16597.74219	
5	3	13116.00293	
6	4	17322.76953	
7	5	14773.04102	
8	6	22700.51172	
9	7	22402.77539	
10	8	23965.1582	
11	9	21120.6582	
12	10	15891.11328	
13	11	15891.11328	
14	12	21190.4043	
15	13	21285.375	
16	14	22702.5918	
17	15	32532.83398	
18	16	39233.81641	
19	17	37323.55859	
20	18	6555.646973	
21	19	6439.51416	
22	20	6281.237305	

Figure 8.3 – Downloaded predictions

The downloaded predictions file can now be joined with the original dataset for further analysis.

The second method for scoring a dataset is via the DataRobot batch prediction **application programming interface** (**API**), which will be discussed in the following section.

Generating prediction explanations

In this section, we will focus on how to generate explanations along with predictions for the scoring dataset. After uploading the scoring dataset (as we discussed in the preceding section), you can now go to the **Understand** tab and then select the **Prediction Explanations** tab, as shown in the following screenshot:

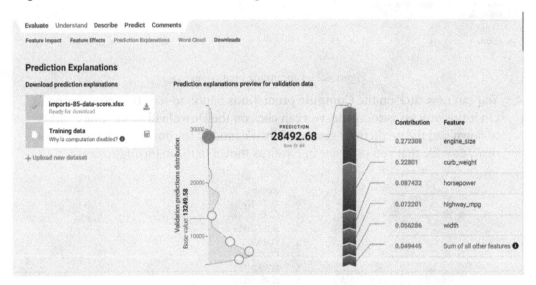

Figure 8.4 – Prediction explanations

In the preceding screenshot, you can see that it now shows the scoring dataset that was uploaded. You can now click on the icon next to the dataset filename to compute the explanations. Once the computation is complete, you will see the download icon. You can use the download icon to download the generated explanations for the predictions made by the model. The explanations come in the form of a `.csv` file that can be opened using Excel, as shown in the following screenshot:

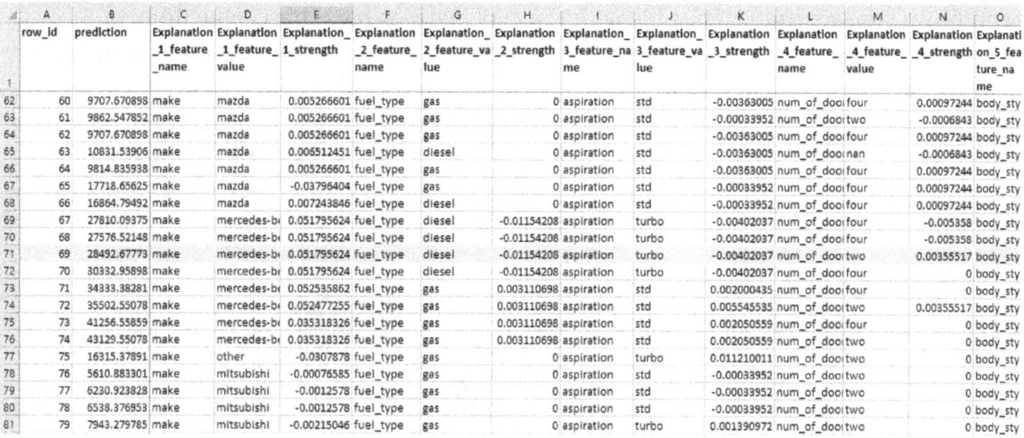

	A	B	C	D	E	F	G	H	I	J	K	L	M	N	O
1	row_id	prediction	Explanation _1_feature _name	Explanation _1_feature_ value	Explanation_ 1_strength	Explanation _2_feature_ name	Explanation _2_feature_va lue	Explanation_ 2_strength	Explanation_ 3_feature_na me	Explanation_ 3_feature_va lue	Explanation _3_strength	Explanation _4_feature_ name	Explanation _4_feature_ value	Explanation _4_strength	Explanati on_5_fea ture_na me
62	60	9707.670898	make	mazda	0.005266601	fuel_type	gas	0	aspiration	std	-0.00363005	num_of_dooi	four	0.00097244	body_sty
63	61	9862.547852	make	mazda	0.005266601	fuel_type	gas	0	aspiration	std	-0.00033952	num_of_dooi	two	-0.0006843	body_sty
64	62	9707.670898	make	mazda	0.005266601	fuel_type	gas	0	aspiration	std	-0.00363005	num_of_dooi	four	0.00097244	body_sty
65	63	10831.53906	make	mazda	0.006512451	fuel_type	diesel	0	aspiration	std	-0.00363005	num_of_dooi	nan	-0.0006843	body_sty
66	64	9814.835938	make	mazda	0.005266601	fuel_type	gas	0	aspiration	std	-0.00363005	num_of_dooi	four	0.00097244	body_sty
67	65	17718.65625	make	mazda	-0.03796404	fuel_type	gas	0	aspiration	std	-0.00033952	num_of_dooi	four	0.00097244	body_sty
68	66	16864.79492	make	mazda	0.007243846	fuel_type	diesel	0	aspiration	std	-0.00033952	num_of_dooi	four	0.00097244	body_sty
69	67	27810.09375	make	mercedes-b	0.051795624	fuel_type	diesel	-0.01154208	aspiration	turbo	-0.00402037	num_of_dooi	four	-0.005358	body_sty
70	68	27576.52148	make	mercedes-b	0.051795624	fuel_type	diesel	-0.01154208	aspiration	turbo	-0.00402037	num_of_dooi	four	-0.005358	body_sty
71	69	28492.67773	make	mercedes-b	0.051795624	fuel_type	diesel	-0.01154208	aspiration	turbo	-0.00402037	num_of_dooi	two	0.00355517	body_sty
72	70	30332.95898	make	mercedes-b	0.051795624	fuel_type	diesel	-0.01154208	aspiration	turbo	-0.00402037	num_of_dooi	four	0	body_sty
73	71	34333.38281	make	mercedes-b	0.052535862	fuel_type	gas	0.003110698	aspiration	std	0.002000435	num_of_dooi	four	0	body_sty
74	72	35502.55078	make	mercedes-b	0.052477255	fuel_type	gas	0.003110698	aspiration	std	0.005545535	num_of_dooi	two	0.00355517	body_sty
75	73	41256.55859	make	mercedes-b	0.035318326	fuel_type	gas	0.003110698	aspiration	std	0.002050559	num_of_dooi	four	0	body_sty
76	74	43129.55078	make	mercedes-b	0.035318326	fuel_type	gas	0.003110698	aspiration	std	0.002050559	num_of_dooi	two	0	body_sty
77	75	16315.37891	make	other	-0.0307878	fuel_type	gas	0	aspiration	turbo	0.011210011	num_of_dooi	two	0	body_sty
78	76	5610.883301	make	mitsubishi	-0.00076585	fuel_type	gas	0	aspiration	std	-0.00033952	num_of_dooi	two	0	body_sty
79	77	6230.923828	make	mitsubishi	-0.0012578	fuel_type	gas	0	aspiration	std	-0.00033952	num_of_dooi	two	0	body_sty
80	78	6538.376953	make	mitsubishi	-0.0012578	fuel_type	gas	0	aspiration	std	-0.00033952	num_of_dooi	two	0	body_sty
81	79	7943.279785	make	mitsubishi	-0.00215046	fuel_type	gas	0	aspiration	turbo	0.001390972	num_of_dooi	two	0	body_sty

Figure 8.5 – Prediction explanations file

In the preceding screenshot, we see that the file contains the predictions, as well as an explanation for each prediction. For example, if we look at row **69** that is highlighted in *Figure 8.5*, we see that the value of make explains 5.17% of the value difference from the base for this automobile. Similarly, you can see the relative contribution of each feature value. Notice that the features in the file are not sorted by the most important feature and also that the most important feature for a given row is not the same as some other row. The feature importance will change from row to row.

Now that we have the predictions and their explanations, let's look at how to analyze these and determine how to use them to take actions or make decisions.

Analyzing predictions and postprocessing

Before we charge off to deploy the model, it would be advisable to analyze the predictions and see if they make sense, whether there are some patterns in the errors, and also how to turn the predictions into something actionable. These are aspects where traditional data science tools and methods are not of much help, and you need to rely on judgment and methods from other disciplines to help formulate the next steps. For this, let's start by combining the scoring dataset file with the explanations file. This can be done in **Structured Query Language** (**SQL**), Python, or Excel. The combined file looks something like this:

	A	V	W	X	Y	Z	AA	AB	AC	AD	AE	AF	AG	AH	AI	AJ
	symboling	compression-ratio	horsepower	peak-rpm	city-mpg	highway-mpg	price	volume	mpg-ratio	cylinder-size	row_id	prediction	ERROR	Explanation_1_feature_name	Explanation_1_feature_value	Explanation_1_strength
2	3	9	111	5000	21	27	13495	528019.9	1.285714	32.5	0	15220.257	-1725.2568	make	other	-0.00461
3	3	9	111	5000	21	27	16500	528019.9	1.285714	32.5	1	15220.257	1279.7432	make	other	-0.00461
4	1	9	154	5000	19	26	16500	587592.6	1.368421	25.33333	2	16642.678	-142.67773	make	other	-0.02181
5	2	10	102	5500	24	30	13950	634817	1.25	27.25	3	13325.879	624.12109	make	audi	0.002237
6	2	8	115	5500	18	22	17450	636734.8	1.222222	27.2	4	17775.49	-325.49023	make	audi	0.013684
7	2	8.5	110	5500	19	25	15250	624190	1.315789	27.2	5	14353.275	896.72461	make	audi	0.008239
8	1	8.5	110	5500	19	25	17710	766364	1.315789	27.2	6	18259.832	-549.83203	make	audi	0.017086
9	1	8.5	110	5500	19	25	18920	766364	1.315789	27.2	7	18972.904	-52.904297	make	audi	0.017086
10	1	8.3	140	5500	17	20	23875	769115.8	1.176471	26.2	8	23977.549	-102.54883	make	audi	0.058189
11	0	7	160	5500	16	22 ?		629188.6	1.375	26.2	9	22835.09	#VALUE!	make	audi	0.059683
12	2	8.8	101	5800	23	29	16430	622095.6	1.26087	27	10	16223.387	206.61328	make	bmw	0.004728
13	0	8.8	101	5800	23	29	16925	622095.6	1.26087	27	11	16164.808	760.19238	make	bmw	0.004728
14	0	9	121	4250	21	28	20970	622095.6	1.333333	27.33333	12	21256.549	-286.54883	make	bmw	0.05259
15	0	9	121	4250	21	28	21105	622095.6	1.333333	27.33333	13	21256.549	-151.54883	make	bmw	0.05259
16	1	9	121	4250	20	25	24565	704276.4	1.25	27.33333	14	22461.092	2103.9082	make	bmw	0.056752
17	0	8	182	5400	16	22	30760	704276.4	1.375	34.83333	15	32630.617	-1870.6172	make	bmw	0.028246
18	0	8	182	5400	16	22	41315	706639.4	1.375	34.83333	16	39603.613	1711.3867	make	bmw	0.033169

Figure 8.6 – Combined scoring data and predictions

We also created a new **ERROR** column that simply subtracts **prediction** from **price**. We can now use Excel to create a pivot table and look at the results from multiple perspectives. For example, let's create a pivot table and look at the **Average of ERROR** value by **symboling**, as shown in the following screenshot:

symboling	Average of ERROR
-2	-217.05
-1	15.90
0	-10.37
1	55.23
2	90.86
3	2.94

Figure 8.7 – Average of ERROR value by symboling

The preceding screenshot shows that errors are much higher for the value **-2**. Looking at the dataset, we find that we have only three data points for **-2**, thus it is not a surprise that the model performs poorly. This tells us that we cannot trust the results when the **symboling** value is **-2** and that we should try to get more data for this value. Analysis such as this can point to areas of improvement and where to focus your efforts. We also realize that since this is an average error, we should use the average of the absolute percentage value of the error to prevent incorrect conclusions, as shown in the following screenshot:

symboling	Average of abs perc ERROR	Count of abs perc ERROR2
-2	0.054	3
-1	0.041	22
0	0.050	65
1	0.034	52
2	0.037	32
3	0.036	27

Figure 8.8 – Average of abs perc ERROR value by symboling

Now, we see that the absolute percent error decreases as the **symboling** value increases. At this point, there is no hard and fast way to find insights except exploring the output data and looking at it from different perspectives to see what you can find. Typically, it is a good idea to sort the errors and look at rows that have unusually large errors, and then see if you can determine why this is so.

Now, on to one of the most important aspects of building a data science model— understanding which actions to take. Now that we have a reasonable model to predict price, a question arises: *What should we do with this information?* Hopefully, the answer was determined at the start of the project as to what was the goal of this exercise. Let's assume that the objective is to set the price of a new vehicle by looking at the prediction of the model and providing all the parameters such as engine_size, and so on. We could also imagine that a model such as this could be useful even during the design stage when designers are trying to determine trade-offs between different parameters such as bore or width. This goes on to say that a predictive model can many times be applied to use cases that were not considered while building the model.

This, however, requires us to understand the broader context of the business problem. This is the primary reason we took time to discuss and understand the business context in *Chapter 3, Understanding and Defining Business Problems*. It might be useful to revisit that chapter to refresh the concepts discussed there as we will use some of the techniques that were introduced there, such as causal modeling.

To determine how we use price prediction, let's review what we know about how price relates to other parameters. In *Chapter 5, Exploratory Data Analysis with DataRobot,* we looked at association analysis information. Association strengths using mutual information were generated by DataRobot. We can use that information to draw a network graph between different features, as shown in the following screenshot. You can do this by drawing a circle for each feature, and then creating lines between features that have high association strengths:

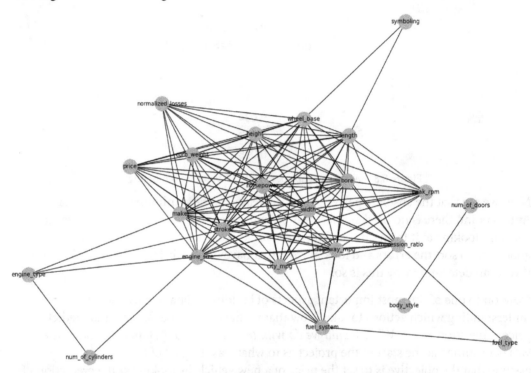

Figure 8.9 – Network graph of associations between features

In *Chapter 7, Model Understanding and Explainability,* we saw the feature importance for price in terms of **SHapley Additive exPlanations (SHAP)** values is specific to the model we selected. The following might represent a causal diagram for this problem:

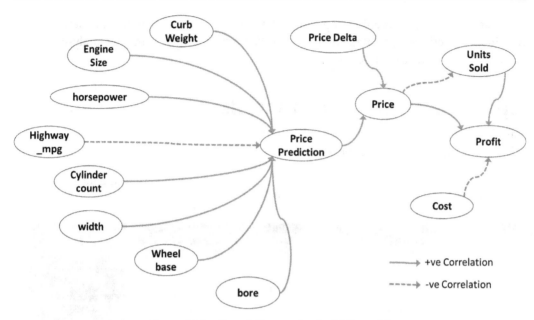

Figure 8.10 – Causal diagram for the XGB model

The left side of the diagram represents the most important features from the SHAP values. Let's imagine that the actual price charged is a bit different from the prediction. The **Price Delta** feature reflects a decision someone might make to charge a price different from the prediction. The **Price** feature impacts **Units Sold**, which ultimately affects the profitability. Note that this reflects just one possible way of using this model to help make pricing decisions.

If, on the other hand, we imagine that we are trying to help the car design team come up with the best car configuration that will also be the most profitable one, then we might look at the diagram a bit differently. This is because different choices of car or engine design will also impact the cost of the car. Also, we know from *Figure 8.9* that the features are not independent. Changing the **bore** feature will change the **Engine Size** and the **Horsepower** features. Hence, when we are looking into making decisions, we have to think about the causal impacts as well. This is a very simplified view, and you can imagine that for a real problem, these diagrams will be a lot more complex. Imagine business leaders making those decisions by taking into account all of these relationships in their heads. This is one of the reasons that many times, models are not used by business users.

In our example problem, the causal diagram shown in *Figure 8.10* is fairly simple. You can imagine real-world problems where this diagram will be a lot more complex. In such cases, it is very difficult to assess the impact the deployment of a model will have on the ecosystem. This includes users and other stakeholders. Complex problems tend to have many unanticipated consequences, especially when the affected parties are people.

In such situations, if the potential impact may be large, it is advisable to test the new model in a synthetic or simulated environment. With the testing and impact analyses complete, we are now ready to deploy our model.

Deploying DataRobot models

DataRobot makes it pretty easy to deploy the models you have developed. To prepare a model for deployment, here are the steps:

1. Let's unlock the project so that we can see the metrics for the holdout datasets, as shown in the following screenshot:

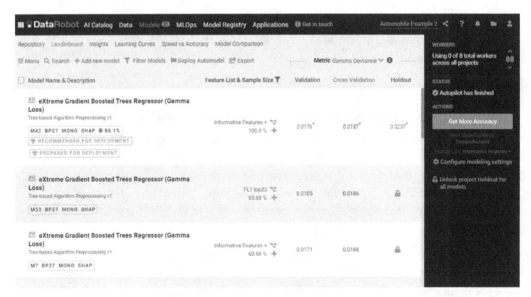

Figure 8.11 – Unlocking DataRobot models

In the preceding screenshot, you can see the **Unlock project Holdout for all models** option on the right side of the interface.

2. You should unlock the project only after you have selected the model that you are choosing for deployment. In our case, we have selected the XGB model that uses the **FL1 top23** feature list. Clicking on this option brings up a dialog box, as shown in the following screenshot:

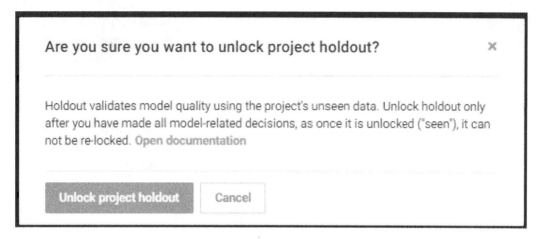

Figure 8.12 – Unlocking project holdout

3. Unlocking the project is an irreversible process. Let's unlock the project and see the holdout metrics, as shown in the following screenshot:

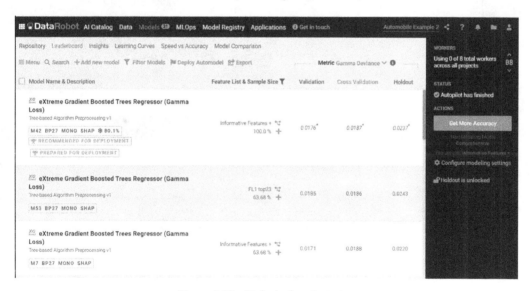

Figure 8.13 – Unlocked project view

Figure 8.13 shows that the holdout values are higher than the cross-validation values, as expected. The holdout values are a better representation of the kind of performance you should expect from a model after deployment.

4. Now that the project is unlocked, let's retrain the selected model with 100% of the data to improve this model's performance. For that, click on the orange + sign for the model, as shown in *Figure 8.13*. This will bring up a dialog box for changing the sample size, as shown in the following screenshot:

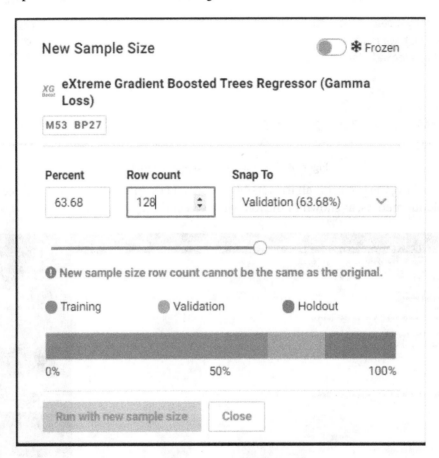

Figure 8.14 – Defining new sample size

In the preceding screenshot, you see options to change the sample size.

5. Drag the slider bar all the way to **100%** to indicate that you want to train the model with 100% of the data, as shown in the following screenshot:

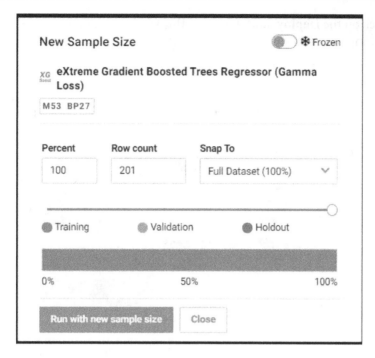

Figure 8.15 – Setting new sample size

6. You can now click the **Run with new sample size** button. DataRobot will now retrain the XGB model with 100% of the data. For the XGB model, you can now click on the **Predict** tab and then the **Deploy** tab, as shown in the following screenshot:

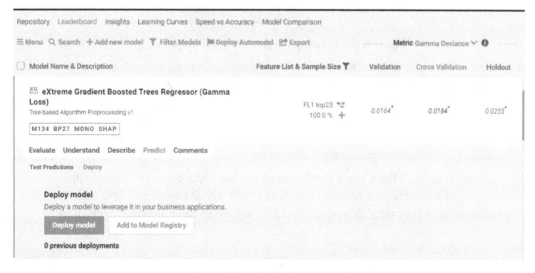

Figure 8.16 – Deploying a model

7. Next, click on the **Deploy model** button. This will bring up a new page, as shown in the following screenshot:

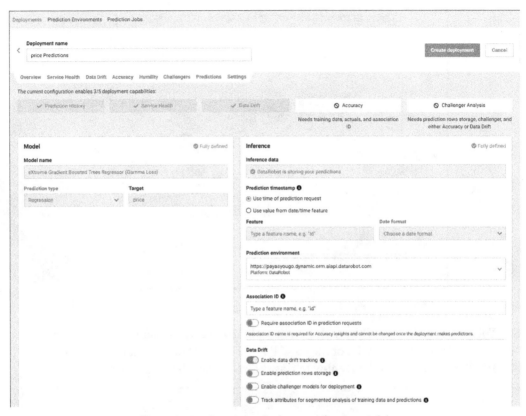

Figure 8.17 – Creating a deployment for the model

8. You can now give a name to your deployed model. You can also select your prediction environment where the deployed model is hosted, as set up by your administrator. Under the **Data Drift** section, you can specify if you want to track data drift or enable challenger models. You can also enable the storage of prediction rows, which allows DataRobot to analyze performance over time. Similarly, you can enable the tracking of attributes for segment-based analysis of model performance.

9. You can now click the **Create deployment** button. DataRobot will now deploy your model and create a baseline for model drift. Once completed, you will see information about your deployed model, as shown in the following screenshot:

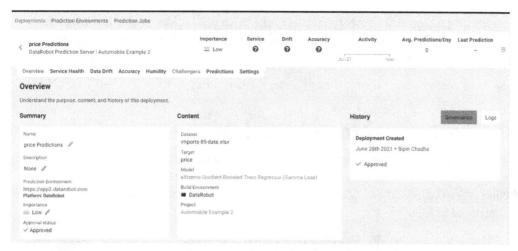

Figure 8.18 – Deployed model overview

You can now see the endpoint for the **REpresentational State Transfer (REST)** API for your prediction model. For example, for the **price Predictions** model, the **prediction environment** is `https://app2.datarobot.com`.

10. You can now invoke this API to generate predictions. You can also see other information about your deployment by clicking on different tabs. If you click on the **Service Health** tab, you will see a page like this:

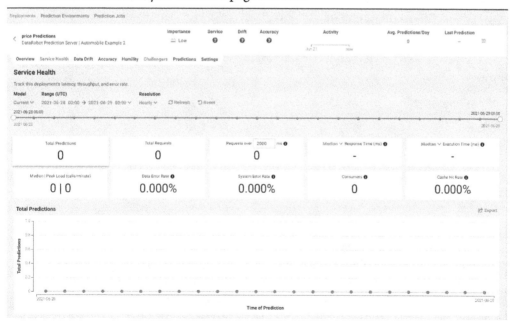

Figure 8.19 – Service health of deployments

The preceding screenshot shows the status of the price prediction model. It shows how many predictions have been done, the response time for a prediction, and the error rates. The screenshot does not show any values because we just deployed this model.

We are now ready to start monitoring this deployed model.

Monitoring deployed models

As you will have guessed by now, the job of the data science team does not end once a model is deployed. We now have to monitor this model to see how it is performing, whether it is working as intended, and if we need to intervene and make any changes. We'll proceed as follows:

1. To see how that works, let's click on the **Predictions** tab, as shown in the following screenshot:

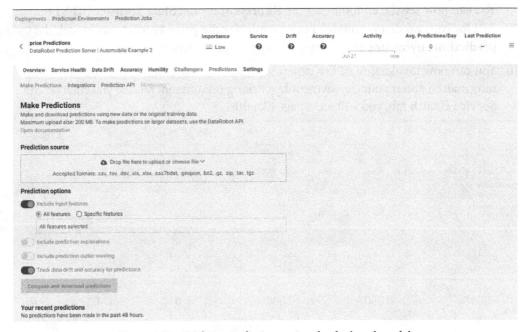

Figure 8.20 – Making predictions using the deployed model

2. We can now upload a dataset to be scored, by dragging and dropping a file (here, we will use the same file that we used before during model training) into the **Prediction source** box. We can now see other options becoming available, as shown in the following screenshot:

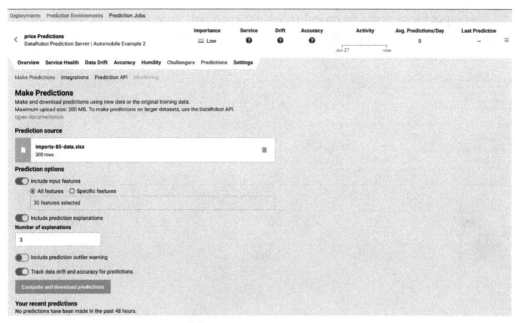

Figure 8.21 – Computing predictions for a dataset

3. After selecting the options, we can click on the **Compute and download predictions** button. After DataRobot finishes the computations, we will see the output file becoming available, as shown in the following screenshot:

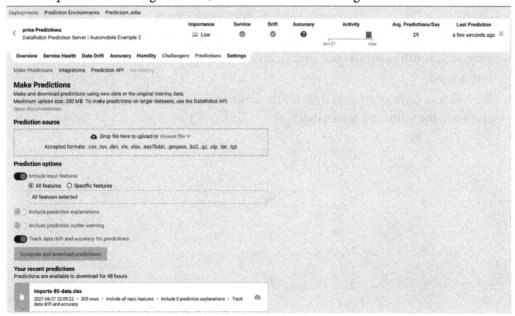

Figure 8.22 – Downloading predictions

The output file can now be downloaded and analyzed. Since we are interested in monitoring the model, let's click on the **Service Health** tab, as shown in the following screenshot:

Figure 8.23 – Service health of the model

We can now see that the model has serviced 15 requests with a median response time of 325 **milliseconds (ms)** and an error rate of 0%. The overall service health looks good.

4. We can now look at the data drift for the model by clicking on the **Data Drift** tab, as shown in the following screenshot:

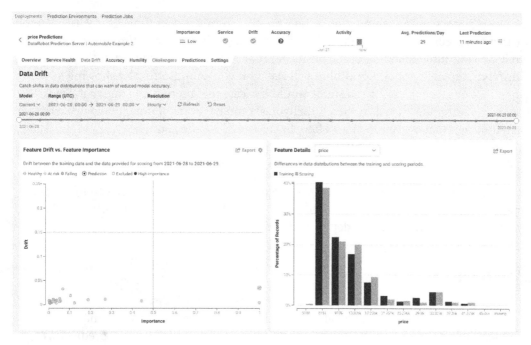

Figure 8.24 – Data drift for the model

In the preceding screenshot, at the top of the **Data Drift** page, we see the data drift between the scoring data and training data. The left graph shows drift by feature importance, and we can see that the amount of drift is very low. This is not surprising since we used the same dataset. For real datasets, the drift will be a bit higher. Similarly, the graph on the right shows the distribution of records grouped by price. Here again, we see that the distributions are very similar for the target feature, `price`. If you scroll down the page, you will see additional graphs, as shown in the following screenshot:

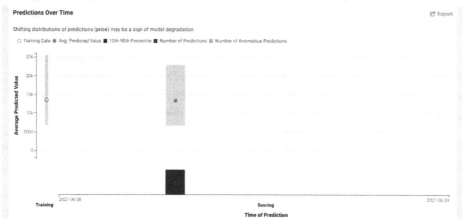

Figure 8.25 – Data drift for the model: additional information

The preceding screenshot shows the average prediction values over time. This will indicate whether the predictions have been stable or if they have changed over time. You will have to rely on your understanding of the business problem to determine whether the amount of drift is acceptable or not. DataRobot will also give you an indication by showing a red, yellow, or green status. A red status would indicate that there is an issue that needs to be resolved; similarly, yellow means that you should be aware of potential issues, and green indicates that everything looks fine. In general, the issue could be errors in the data pipeline or a change in the business environment. A change in the business environment would be an indication that the model needs to be retrained.

If the model needs to be retrained or if you need to rebuild the model, you can follow the steps that we have outlined in the preceding chapters. This completes a basic view of how you use DataRobot to build and deploy a model.

Summary

In this chapter, we learned how to use models after training. We discussed the methods that are used to score a dataset and also methods that are used for analyzing the resulting outputs. We also covered methods and considerations for turning predictions into actions or decisions. This is a critical step whereby you have to engage with your business stakeholders to make sure that introducing this model will not cause unforeseen problems. This is also the time to work on change management tasks such as communicating changes to people who are impacted by the change and ensure that users are trained in the new process and know how to use the new capabilities.

We then discussed how to use DataRobot capabilities to rapidly deploy a model and then monitor the model performance. It is easy to underestimate the importance of this capability. Model deployment and monitoring are not easy, and many organizations spend a lot of time and effort trying to deploy a model. Hopefully, we have shown how easily this can be accomplished with DataRobot.

We have now completed the basic steps needed to build and deploy a model and can now go over some advanced concepts and capabilities of DataRobot. You are now ready to dive into advanced topics based on your interest or based on the type of project you will be working on. For example, if you are working on a time series problem, then you can review *Chapter 9, Forecasting and Time Series Modeling*.

Section 3: Advanced Topics

This section covers many of the advanced topics and capabilities that you can leverage once you have mastered the previous sections. This section provides examples of advanced capabilities that experienced data scientists will use to make them more productive. Some chapters in this section require a familiarity with Python programming.

This section comprises the following topics:

- *Chapter 9, Forecasting and Time Series Modeling*
- *Chapter 10, Recommender Systems*
- *Chapter 11, Working with Geospatial Data, NLP, and Image Processing*
- *Chapter 12, DataRobot Python API*
- *Chapter 13, Model Governance and MLOps*
- *Chapter 14, Conclusion*

9
Forecasting and Time Series Modeling

In this chapter, we will understand what time series are and will see how DataRobot can be used to model them. Time series modeling is becoming increasingly useful in businesses. However, the challenges associated with forecasting make it quite challenging for many skilled data scientists to successfully carry out time series modeling, and this form of modeling could also be extremely time-consuming. DataRobot provides an automated process that enables data scientists to carry out time series projects in an effective and efficient fashion. In this chapter, we will introduce the concept of forecasting, stressing its commercial importance and inherent challenges, and illustrate how DataRobot can be used to build its models.

By the end of this chapter, you will have learned how to utilize DataRobot in building time series forecasting models. In addition, we will look at making predictions with these models. We go further by building models for multi-series time series as part of the advanced topics. Here are the main topics to be covered in this chapter:

- Conceptual introduction to time series forecasting and modeling
- Defining and setting up time series projects
- Building time series forecasting models and understanding their model outcomes
- Making predictions with time series models
- Advanced topics in time series modeling

Technical requirements

Some parts of this chapter require access to the DataRobot software and some tools for data manipulation. Most of the examples deal with small datasets and therefore can be handled via Excel. The dataset that we will be using in this chapter is described next.

Appliances energy prediction dataset

This dataset can be accessed at the *University of California Irvine* (*UCI*) Machine Learning Repository (`https://archive.ics.uci.edu/ml/datasets/Appliances+energy+prediction#`).

> **Dataset citation**
>
> *Luis M. Candanedo, Véronique Feldheim, Dominique Deramaix, Data driven prediction models of energy use of appliances in a low-energy house, Energy and Buildings, Volume 140, 1 April 2017, Pages 81-97, ISSN 0378-7788.*

This dataset captures temperature and humidity in various rooms in a house and in the outside environment, along with energy consumption by various devices over time. The data is captured every 10 minutes. This is a typical example of a time series dataset. Data is provided in .csv format and the site also provides descriptions of the various features. All features in this dataset are numeric features. The dataset also includes two random variables to make the problem interesting.

Conceptual introduction to time series forecasting modeling

The dynamic nature of the commercial environment makes time a pivot resource for business success. As a result, businesses need to account for the time factor in their decision-making. Changes occur within commercial settings at a high pace, which makes it pertinent for organizations to take rapid yet considered actions. Analytic technology provides organizations with tools that enable forecasting of the future so that decision-makers have crucial time in hand to ensure their decision aligns with their organizational objectives. Organizations use time-specific data to predict the volume of sales in a future period. Other writers have differentiated time series modeling from forecasting models. In this chapter, we have used the term interchangeably and consider **time series forecasting** to involve the use of advanced analytics to gain insights that guide business decisions leveraging time-based data.

Time series forecasting supports numerous aspects of business planning. With forecasting, human and other forms of resource planning can be optimized to ensure that expected outcomes are realized. Through forecasting, cash flow, profit, and budgeting projections are more rigorously established, thereby mitigating human bias. Forecasting sales could be influenced by several factors that are controllable and non-controllable. Certain consumer factors that change with time tend to affect the volume of sales. These factors include changes in population, customer taste, and interests. In addition, demand is sensitive to broader economic variables, such as inflation, that also change with time. As a result, it becomes pertinent to use some features that could act as proxies for these consumer and economic variables in addition to **lagged** or historic sales. Because some of these variables are challenging to acquire, analysts tend to be limited to a few historic values and volumes in modeling future outcomes.

Although a detailed discussion on time series is out of the scope of this book, it is, however, pertinent to appreciate that the properties of modeling time series make them more challenging to work with. In addition to difficulties with other forms of predictive modeling discussed in previous chapters of this book, time series modeling comes with additional challenges. One of the assumptions of linear regression modeling is that of independence of observations, that is, that observations or data rows are independent. However, this assumption is inevitably broken with time series modeling. Within time series, **autocorrelation** occurs naturally, as observations are similar across different time periods. It is also possible that highly corrected observations don't occur successively, in which case **seasonality** occurs. Series are considered seasonal when observations across a fixed time frame have higher levels of correlation. Indeed, these are periodic fluctuations in observations. A similar volume of sales of flight tickets during holiday periods brings this to life. Seasonality could indeed occur yet fails to follow a fixed time frame, described as **cyclicity**. Qualifying cycles generally require considerably larger datasets than other properties of series as cyclicity is mostly related to external factors such as macroeconomic or political changes within the business environment.

Autocorrelation also gives rise to **linearity**, a concept that describes an overarching trend where consecutive observations are similar, albeit changing in such a way that they follow a linear trend. Due to this linear trend, albeit sometimes with some integrated fluctuations, the mean of specific time frames will follow a pattern but is unlikely to be the same, hence the use of **moving average** (**MA**) and **autoregression** approaches to represent time series. However, series can still be characterized by the extent to which their statistical properties change over time. They are considered **stationary** when they have a constant mean and variance that are independent of time. What is most interesting, albeit problematic statistically, is that some time series data has a combination of these properties. A good example is the volume of flights. Though gradually increasing over time, being seasonal, during an economic downturn this falls generally. In this example, we can see elements of seasonality, cyclicity, and linearity.

Another concept that sometimes gets lost in the details is that of **actionability**. Actionability being the ability of stakeholders to act because of an analysis or a model's outcome, it is very common for data scientists to focus on the accuracy of predictions. While accuracy is important, what is more important is to provide actionable guidance to decision-makers. A forecast that enables you to take action today is more valuable than a forecast that is more accurate but not actionable. Care must be taken while defining the forecasting problem to ensure the actionability of the model being developed.

The foregone conversation in this section highlights the properties that make time series modeling more challenging for typical data scientists. DataRobot has developed unique processes that enable data scientists, including those with limited statistical exposure, to create complex yet robust time series models. In the subsequent section, we will look at how to define and set up time series problems in DataRobot.

Defining and setting up time series projects

In *Chapter 4, Preparing Data for DataRobot*, through to *Chapter 8, Model Scoring and Deployment*, we explored the creation, understanding, scoring, and deployment of basic models in DataRobot. We saw that DataRobot automatically built several models for us and we could then score a dataset using these built models. Further, after we have chosen a model that best aligns with our needs, DataRobot provides us a process to deploy our selected model. Due to the difference between time series modeling and other forms of predictive modeling, we will explore in this section how to mitigate problems by effectively defining and setting up time series projects in DataRobot.

The dataset we will use to explore the use of time series modeling with DataRobot is the Appliances energy prediction dataset that we explored in *Chapter 4, Preparing Data for DataRobot*. The goal of the project is to predict energy usage. This energy usage time series dataset has 4 and a half months' worth of 10-minute readings from differing data sources. First, the data involved room temperature and humidity in a house. These were monitored using a wireless sensor network and the data was stored every 10 minutes. Each of the nine rooms in the house had their readings for temperature and humidity stored for the time frame. Second, there was external data that provided a nearby airport (public source) detailed information pertaining to weather information outside the house, again with a 10-minute interval. This included wind speed, visibility, dew point, pressure, and humidity. This information was merged with the data using date and time. In addition, appliances and light usage aligned to date and time were attached to the dataset.

Within this dataset, it is easy to see that the goal of this time series prediction is predicting energy usage. The immediate influencing variables are the temperature and atmospheric pressure within the house; however, the external data from the weather outside the house is important. We created features calculating the average conditions across the nine rooms in the house. In addition, we engineered features that captured the difference between the mean room and the external temperature, as well as the difference between the mean room and external pressure. Since we have two time series (appliance usage and light usage), we will approach this problem in two ways. First, as a **single time series**, we will look at the sum of both appliance and light usage. Subsequently, within the advanced section, we will examine the **multiple time series** approach, with which we will be making predictions for each usage type. As with other prediction projects on DataRobot, we ingest the data as a `.csv` file, as seen in the following screenshot:

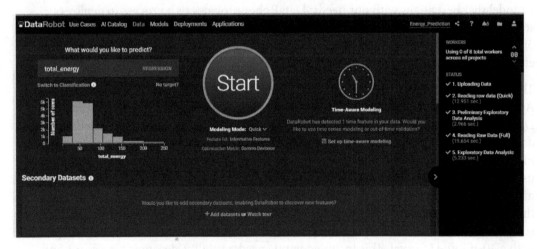

Figure 9.1 – Choosing a target variable for time series

The project is named `Energy_Prediction` and the target variable selected is `total_energy` (the sum of light and appliance usage). We proceed as follows:

1. After selecting a target variable, we select a time variable and the nature of the time-based modeling. Clicking the **Set up time-aware modeling** button, as shown in *Figure 9.1*, highlights the importance of time as a dimension and provides an opportunity to choose a time variable. In this case, we choose the `date` feature, which specifies the date and times of all readings, as illustrated in the following screenshot:

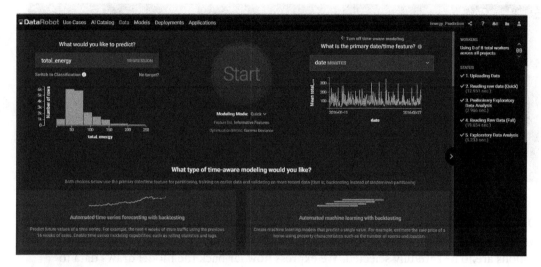

Figure 9.2 – Choosing a time-aware function and time variable

2. Once the **Set up time-aware modeling** button is clicked and the time feature is selected, the platform requests the type of time-awareness model to be built. There are two options—**Automated time series forecasting with backtesting** and **Automated machine learning with backtesting**, as described next:

*   **Automated time series forecasting with backtesting**—This option considers previous data in predicting future data. With time series, there is a need to forecast multiple future points. A case in point for this type of time-aware project could be estimating departmental stores' daily sales for the next month using data from their last year's sales.

Automated machine learning with backtesting—The automated machine learning option, sometimes referred to as **out-of-time validation**, basically creates time-based features in a row and then uses a typical predictive model that predicts a target variable for that row. Here, we do not use the typical cross-validation scheme; instead, this approach employs older data for training and holds back newer data for backtesting. Our project's context problem falls within the forecasting category type, so this option is selected, as seen in the following screenshot:

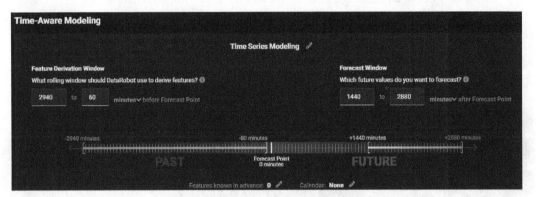

Figure 9.3 – Time-aware modeling options

Once we have selected the **Automated time series forecasting with backtesting** option, we are presented with a **Time-Aware Modeling** options tab (see *Figure 9.3*). Here, a few options are to be carefully selected. We express how far back the model draws data to make predictions and also how far forward the model makes predictions for. Let's first consider the **Feature Derivation Window** option. This **rolling window** highlights a lag upon which features and statistics for time series models are derived in relation to the time from which a forecast is made (**forecast point**). The rolling window is expressed in relation to the forecast point and automatically moves forward with the passage of time. In an ideal situation, this window should cover a seasonal period in your data. Essentially, this window typically answers the question: *How far back does the data our model uses to make predictions stretch?* Also, there should be enough time between the end of the window and your forecasting time to cater for any data ingest delays still limiting this time gap, ensuring the data is recent enough. This period is known as the **blind history**. In our case, we have assumed that an hour would be enough time to allow any blind history, so set the gap before the forecasting point to 60 minutes. Considering our data is limited to 4 and a half months, seasonality within the context of our problem would be day and night usage. Accordingly, we have set our rolling window to 2 days (2,880 minutes), which, when accounting for the initial 60-minute forecast point gap, amounts to 2,940 minute

The second consideration is for the **Forecast Window** option. This defines, in relation to the forecast point, how far in the future we are predicting. This has two elements; first, when the prediction starts. The predictions should provide enough time for actions to be taken yet not be too far in the future to ensure these predictions are accurate enough. Secondly, we select our prediction end. This is dependent on the start point as well as the nature of our problem. So, this aspect answers the question: *How far forward should predictions be made?* For the problem at hand, we have selected an **operationalization gap**, a gap between the forecast point and the start of the prediction window of 1 day (1,440 minutes). Also, the rolling window is set at 1 day, which in consideration of our operationalization gap becomes 2,880 minutes.

Having set up the time series forecasting project in this section, we will now explore the processes around building the models, from understanding feature lists and their distributions to looking at their impacts on evaluating models.

Building time series forecasting models and understanding their model outcomes

Similar to projects we looked at in *Chapter 4*, *Preparing Data for DataRobot*, through to *Chapter 8*, *Model Scoring and Deployment*, once we have finished with the initial configurations, we scroll up and click on the **Start** button. By doing this, DataRobot automatically builds time series models for this project. Before we evaluate the models, it would be useful to understand the nature of the features the platform extracts. DataRobot extracts features from the data that differ considerably from those of other prediction models, as is evident in the following screenshot:

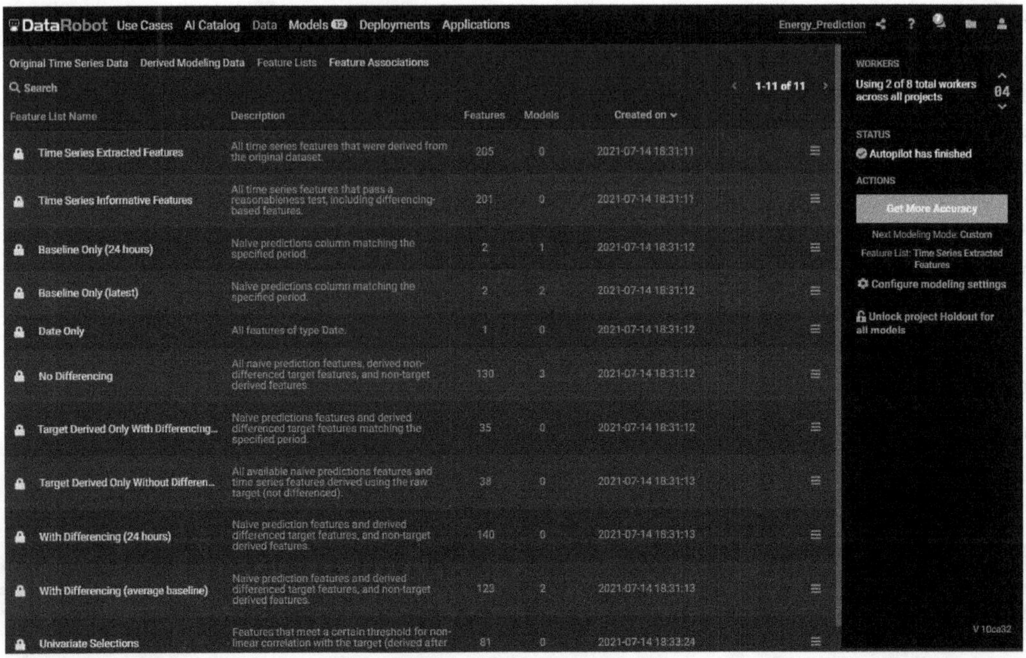

Figure 9.4 – Feature lists

The lists shown under the **Feature Lists** tab are constructed as part of **exploratory data analysis (EDA)** and itemize differing lists of features that DataRobot employs in creating models. Many of the feature lists involve **derived features**, which are created automatically based on properties of time series. A further discussion on derived features will be carried out later in this section. It is easy to see that some of the lists involve features that are extracted from the original data (for example, **Time Series Extracted Features**). Others involve features created solely from dates, while some are assessed as informative. Most lists appear to be combinations of differing types (for example, **Time Series Informative Features**). Importantly, the feature lists provide the descriptions as well as the number of features for each feature list name. Feature lists that could be pivotal are presented as part of the **Leaderboard** feature, as illustrated in the following screenshot, which guides our final model choice:

Figure 9.5 – Model leaderboard

The **Leaderboard** feature offers insights into models that have been built for a DataRobot project. It provides information regarding model names and **identifiers** (**IDs**), their accuracy metrics, and their types, versions, and sample sizes for the model development. With time series modeling, however, there are some differences, as noted next. Firstly, the sample size is present in data ranges. This is due to the time-based nature of time series datasets. Unlike other modeling forms, the time order of the data does affect outcomes; as a result, data is selected in time ranges. In this case, as can be seen in *Figure 9.5*, our models were built using 3 months', 21 hours', and 51 minutes' worth of data. Secondly, instead of the **Validation** and **Cross-Validation** columns, we have the **Backtest 1** and **All Backtests** columns. The backtests follow logically from the discussion regarding the sample size (see *Chapter 6, Model Building with DataRobot*). The backtests provide an evaluation of the model performance on a subset of the data. However, unlike a typical validation, the data is time-ordered, and the size and number of backtests can be altered as needed. We have used the default backtest setting for this example project so that the data was partitioned in such a way that only one backtest partition was available for modeling. Finally, with time series modeling projects, there appear to be more feature lists. As with other predictive project types, the models could be ordered or selected using any of the columns on the **Leaderboard** feature.

There are a number of metrics against which time series forecasting models could be assessed. This, of course, depends on the model. For regression-type outcomes, some advocate the use of **Root Mean Square Error** (**RMSE**). The nature of the problem remains critical in determining the metrics for assessment. That said, the role of the **baseline model** on the leaderboard is crucial to evaluating other models. The baseline model employs the most recent value in making its predictions. As such, comparing models with the baseline prediction blueprint plays a pivotal role in the model evaluation as it somewhat answers the question: *To what extent are our models better than a naïve prediction from the most recent data?* DataRobot provides the **Mean Absolute Scaled Error** (**MASE**), which compares the **Mean Absolute Error** (**MAE**) of models of interest with those of the baseline model. For instance, the **Eureqa Generalized Additive Model (250 Generations)** model, as presented in the following screenshot, has a comparative ratio of about 0.76 for **Backtest 1**. This suggests that the Eureqa model is about 24% better than the baseline. Since the **Holdout** metric could highlight considerable changes within the data, it should be included in model evaluation but not used in isolation. Other indications when evaluating models are covered within the *Advanced topics in time series modeling* section of this chapter. Model names could be clicked to provide elaborate insights about the data and its processes. We now turn to those we consider unique to time series forecasting, using the **Eureqa Generalized Additive Model (250 Generations)** example here:

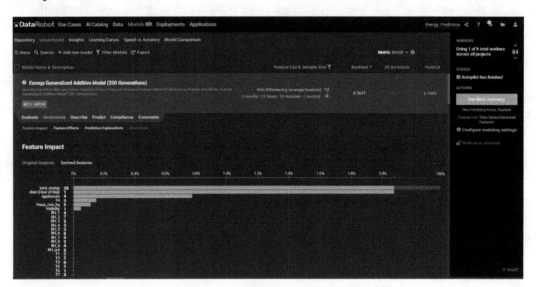

Figure 9.6 – Impact of original features

The **Understand** tab presents us with **Feature Impact, Feature Effects, Prediction Explanations**, and **Word Cloud** capabilities, which we have already encountered in *Chapter 7, Model Understanding and Explainability*. **Feature Impact** shows the relative extent to which features contribute to a model's overall accuracy. A click on the **Feature Impact** tab opens the **Original features** page (see *Figure 9.6*). The original features are features as they were in the dataset.

The other tab within **Feature Impact** depicts the effect of derived features on the accuracy of the model. As alluded to earlier, derived features are those constructed based on the characteristics of time series. For instance, the stationary nature of some time series suggests that their statistical properties do not change over time. In the case of our model, the most impactful derived feature (`total_energy (1440 minute average baseline)`) is seen to be a feature constructed based on the stationary nature of the time series, as illustrated in the following screenshot. This is because it highlights the importance of the average 1,440-minute baseline energy on the accuracy of the model:

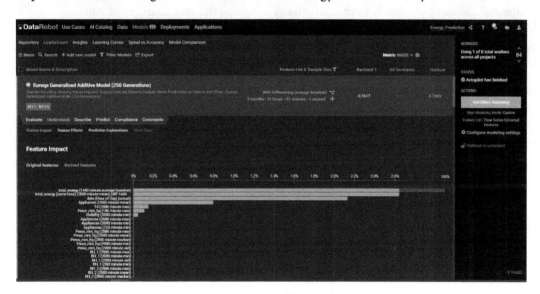

Figure 9.7 – Impact of derived features

It is reasonable, as is evident in *Figure 9.7*, that a considerable number of derived features appear to be created from the stationary property of time series, which on its own could be indicative of this time series being quite stationary. That said, caution needs to be exercised on reaching this conclusion because our dataset only entails 4 and a half months' worth of data; for instance, our dataset only covers January 2016 to May 2016, so does not account for the late Summer, Autumn, and early Winter months. As such, seasonality could occur if we were using a dataset covering a longer time frame.

DataRobot creates features that capitalize on the properties of time series to improve the accuracy of its models. Although not evident in this project, with seasonality or cyclicity, DataRobot establishes when periodic variations occur and creates features accordingly. Based on this information, it next detects patterns of seasonality—for instance, a seasonality that occurs during a time frame could be defined either by counting up from the beginning of the time frame or counting down from the end of the time frame. As such, the platform could detect and build features that, for instance, use energy usage on the last Saturday of March to predict energy usage on the last Saturday of April. In a similar fashion, DataRobot uses features built on **differencing** to improve model performance. It could utilize the average usage during the first week in March as a feature to predict usage during the first week of April.

Moving on to the **Describe** tab, upon opening the **Blueprint** tab, we are exposed to the stages involved in the modeling process of time series projects. As detailed in the following screenshot, we can quickly appreciate that this is not very different from those of other predictive projects encountered in preceding chapters:

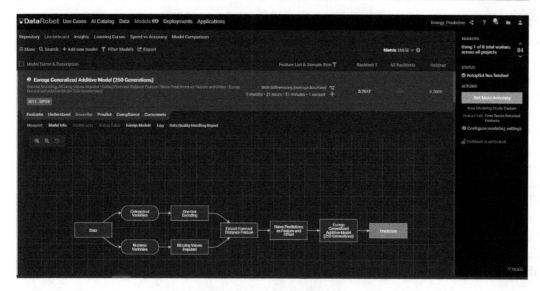

Figure 9.8 – Model blueprint

We have now spent time building and understanding time series forecasting models. The next logical step is to use our selected model to make predictions.

Making predictions with time series models

DataRobot provides us with tools to make predictions pain-free. There are two approaches to making predictions for time series. For small datasets under 1 **gigabyte** (**GB**), predictions could be made using the **Make Predictions** tab on the **Leaderboard** feature. This involves setting up and uploading a prediction dataset, then scoring it within the **Drag and drop a new dataset user interface** (**UI**) functionality. For significantly larger datasets, models need to be deployed and predictions are made using an **application programming interface** (**API**). In this chapter, we will cover the first approach to making predictions. With DataRobot, general model deployments and working with APIs are extensively discussed in *Chapter 12, DataRobot Python API*.

The leaderboard's drag-and-drop approach to scoring models for time series models somewhat differs from those of traditional models, as seen in *Chapter 8, Model Scoring and Deployment*. When the **Make Predictions** tab is opened, DataRobot briefly outlines the recency and quantity of the data needed to make predictions. This outline is mostly consistent with the forecasting windows established as part of the configuration during the model development, as well as features derived. As the prediction process shows in the following screenshot, the prediction dataset requires a minimum of 4,320 minutes of historic data outside of the 60 minutes prior to the forecasting point. In addition, when models include derived features that involve features in earlier time periods, the earlier time period is also included in the dataset requirement. Because the model in question has 24-hours'-difference derived features, this increases the requirement to 5,820 minutes. This 5,820-minute requirement includes an initial 60-minute forecast point gap window, 4,320-minute base prediction requirement data, and 1,440 minutes added on for the derived differencing features. This enables the model to predict 2,880 minutes in advance of the forecasting point after the 1,440-minute operationalization gap. Some of these features are presented here:

Figure 9.9 – Make Predictions window

To make predictions, if the data format is consistent with the training data, proceed
as follows:

1. Click on **Import data**, which allows the data to be ingested from a local source,
    a **Uniform Resource Locator** (**URL**), one of your existing data sources, or AI
    Catalog. If no row is found after the default forecast point, DataRobot generates
    a template. For this to be done, there must be no empty row within the forecast
    window and the template file must meet the upload size limit conditions. After the
    file has been uploaded, DataRobot sets the forecast points and includes the rows
    required to meet the forecast window expectations.

2. Click on the **Compute predictions** button after uploading the data, as illustrated
    in the following screenshot, since the uploaded prediction file is the most recent,
    without gaps and the fill number of rows expected:

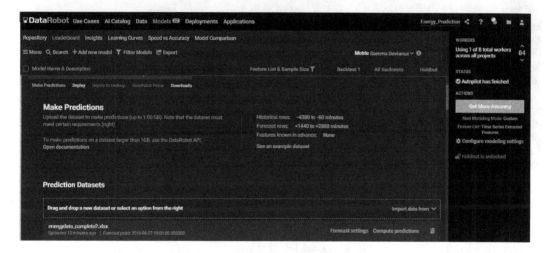

Figure 9.10 – Computing time series predictions

The **Forecast settings** button in *Figure 9.10* provides options for predictions where
either the forecasting point is not expected to be the most recent or changes the
range for which predictions are to be made.

3. To make changes of this nature, click on the **Forecast settings** button, which opens the **Forecast Point Predictions** tab by default, as illustrated in the following screenshot. This window offers a forecast point slide tab selector, which can be configured by either a slide or entering the actual time value. Invalid dates are, however, disabled:

Figure 9.11 – Forecast Point Predictions settings

As alluded to earlier, there is a limit to times that can be selected as a forecast point. The forecast point must be less than or equal to the most recent one. In the case of this project, this is **2016-05-27 19:00:00:00**, which is the most recent data row time, with an operationalization gap of **1440** minutes. A similar operation could be carried out to alter the prediction date ranges. The **Forecast Range Predictions** feature would ideally be used to validate models as opposed to making future predictions.

In this section, we highlighted the importance of ensuring our prediction dataset for time series models is like that for training models. We went on to make predictions and interpreted other outcomes from the model. Next, we will explore more advanced topics involving time series modeling with DataRobot.

Advanced topics in time series modeling

In this chapter, we have learned how to configure, build and make predictions with basic time series forecasting models in DataRobot. In the preceding section, our attention was focused on building models that have one-time series. However, you could have a situation where you might have to make multi-time series predictions. Within the context of our energy utilization problem, we might want to forecast the usage of lights and appliances. Elsewhere, an energy company might want to forecast energy usage for differing cities or households within the same model. We will now take a deep dive into solving problems of this nature. Also, we will explore future ways other advanced approaches may be used in assessing our time series models. Finally, we will acknowledge the role of scheduled events on time series and highlight the provisions made by DataRobot to handle this possibility.

The dataset used for this project highlights the energy usage of lights and other appliances. For the earlier project, we totaled up all usage as our target variable, but in this project (named `Energy_Prediction_2`), models will be built to predict usage for each device type. This dataset now has two series, implying timestamps could recur, yet timestamps within each series must be unique. The differentiating column, `Device_type`, is the ID for the device type that the usage is attributed to. After qualifying the project as being time-aware and choosing its type as **Automated time series forecasting with backtesting** (see *Figure 9.2* for more information on the setup of a time series project), due to the data having multiple rows with the same timestamp, the multiple time series is automatically selected. The next step, as shown in the following screenshot, is to select the series ID, which in this case is `Device_type`:

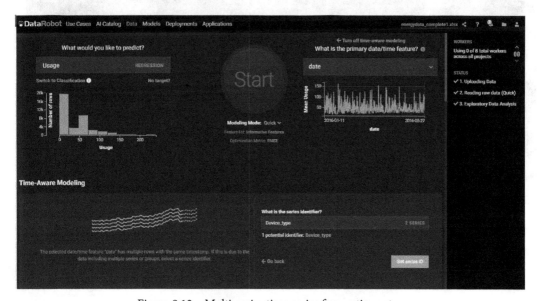

Figure 9.12 – Multi-series time series forecasting setup

For this project, we are interested in further evaluating our models. So, the sequel to customizing our forecasting window, within the **Partitioning** tab of the **Advanced Options** window, is to configure our backtests to help us manage validation folds (for more on validation folds, see *Chapter 6, Model Building with DataRobot*). Here, we simplistically set the number of backtests to 5 + Holdout. The following screenshot details the setup for this configuration, and we can see how the training, validation, and holdout data is partitioned from the initial data. It is important to highlight that to set up the backtests, we must consider any form of seasonality, periodicity, and/or cyclicity within the data and ensure that every fold has at least one instance of these. This is because every backtest should be a complete dataset on its own, so seasonality, periodicity, and cyclicity need to be accounted for within each backtest. The validation and gap lengths can also be altered. The default length for this project is set to over 13 hours and 9 minutes. You can see the configuration here:

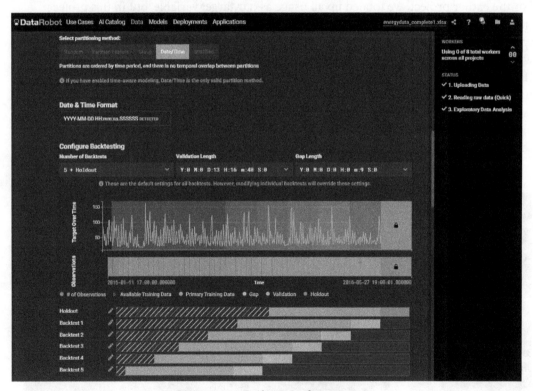

Figure 9.13 – Backtest configuration

Having configured backtesting, we then click on **Start** to train the models. When models are created, the process of evaluation is like that for single time series models. As evident in the following screenshot, we can see the **All Backtests** metric, which measures the average performance of a model across all backtests. As such, it provides an interesting way to quickly assess not only the model performance but also the consistency of the data pattern over time:

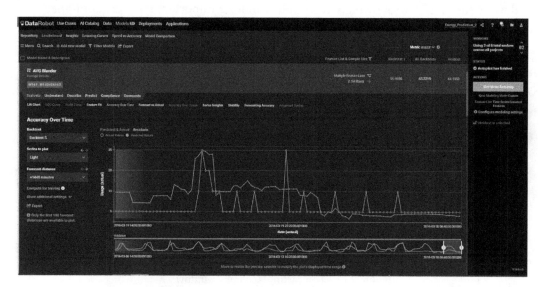

Figure 9.14 – Accuracy over time

The **Accuracy Over Time** feature within the model **Evaluate** tab enables users to have a visual yet in-depth assessment of their models over time (see *Figure 9.14*). Here, the predicted and actual are visually presented. Within this window, you can choose a **Series to plot** setting and alter the **Backtest** and the **Forecast distance** settings. This view, within the context of a business, helps understand if there are periods of poor performance that could imply an aspect of a business not represented in the data. The **Forecasting Accuracy** window, as shown in the following screenshot, is another important representation that suggests how model performance changes as the forecast distance changes:

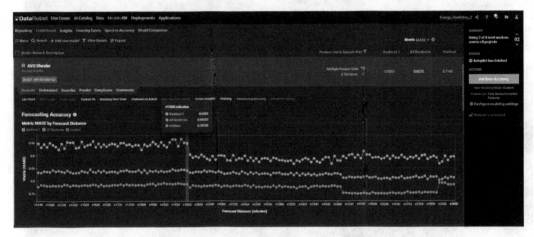

Figure 9.15 – Forecasting Accuracy window

The **Forecasting Accuracy** window highlights alterations in models' performance as forecasts are made into the future. This view allows us to assess where models' performance is similar across time, which is indicative of when models could be used within the business. Furthermore, it highlights when the models' performance considerably exceeds those of the baseline model when the MASE performance metric is used. As illustrated in *Figure 9.15*, the model's performance on **Backtest 1** seems to begin to be considerably better than the baseline model around the +1,960-minute mark. The stability view presents users with the measure of scores across time ranges.

With the quest for better-performing models comes a need to adopt some changes to modeling paradigms. The default models available for time series modeling might just not provide the required performance. In that case, the model repository, as explained in *Chapter 6, Model Building with DataRobot*, presents us with options to select traditional time series models such as **AutoRegressive Integrated Moving Average** (**ARIMA**) and more recent models such as Keras **Long Short-Term Memory** (**LSTM**) and **XGBoost** (**XGB**). Depending on the nature of the time series under investigation, these modeling approaches sometimes present better performance.

Summary

In this chapter, we have extensively examined how DataRobot could be used to build time series models. We briefly discussed the unique opportunities time series modeling presents businesses, as well as the challenges it presents for analysts and data scientists. We used DataRobot to create both single and multiple time series models. We also described how predictions could be made using models built by DataRobot. This was followed by a discussion on advanced aspects of DataRobot's time series capabilities.

Forecasting is extremely important to business because of its ability to foretell what is likely to occur in the future considering time-dependent variables. Another commercially valuable area is the ability to suggest the interest that differing clients would have for a wide array of products. This is where recommender systems come in.

In the next chapter, *Chapter 10, Recommender Systems*, we look at how DataRobot could be used to build recommender engines.

10
Recommender Systems

In this chapter, we will learn about what recommender systems are, discuss their various types, and work through a **DataRobot** implementation of a content-based recommender system. Within this chapter, **recommender system**, **recommendation system**, **recommender engines**, and **recommendation engines** are used interchangeably.

In their simplest form, recommender systems suggest potentially relevant items to users or buyers. In today's commercial environment, businesses tend to have numerous items, products, or services for sale, making it more challenging for users or buyers to connect with their desired products or services. This chapter explains the ubiquity of recommendation engines in the current business space.

Although this book is not the place to cover every aspect of recommendation systems, we will discuss how to utilize DataRobot to build and (make predictions from) recommendation engines and present a conceptual overview of these systems, as well as a brief discussion of their types. Thus, by the end of this chapter, you will learn how to utilize DataRobot to build a content-based recommendation engine. The main topics in this chapter include the following:

- A conceptual introduction to recommender systems
- Approaches to building recommender systems

- Defining and setting up recommender systems in DataRobot
- Building recommender systems in DataRobot
- Making recommender system predictions with DataRobot

Technical requirements

Most parts of this chapter require access to the DataRobot software. The code example is based on a relatively small dataset, Book-Crossing, consisting of three tables, whose manipulation was carried out with **Jupyter Notebook**.

Check out the following video to see the Code in Action at `https://bit.ly/3HxcNUL`.

Book-Crossing dataset

The example used to illustrate the use of DataRobot in building recommendation systems is based on the Book-Crossing dataset by Cai-Nicolas Ziegler and colleagues. This dataset was accessed at `http://www2.informatik.uni-freiburg.de/~cziegler/BX/`.

> **Note**
>
> Before using this dataset, the authors of this book have informed the owner of the dataset about its use in this book.
>
> Cai-Nicolas Ziegler, Sean M. McNee, Joseph A. Konstan, Georg Lausen (2005). *Improving Recommendation Lists Through Topic Diversification. Proceedings of the 14th International World Wide Web Conference (WWW '05).* May 10 – 14, 2005, Chiba, Japan.

The data was collected during a four-week collection of the Book-Crossing community between August and September 2004. The subsequent three tables, provided in CSV format, make up this dataset.

- **Users**: This table presents the profile of the users, with an anonymized `User-ID` presented as integers. Also provided are the users' `Location` and `Age` values.

- **Books**: This table contains the characteristics of the books. Its features include `ISBM`, `Book-Title`, `Book-Author`, `Year-Of-Publication`, `Publisher`.

- **Ratings**: This table shows the book ratings. Each row provides a user's rating for a book. The `Book-Rating` value is either implicit as `0` or explicit between `1` and `10` (the higher the number, the better the rating). However, within the context of this project, we will focus solely on ratings that are explicit for the model development. The table also includes the `User-ID` and `ISBN` values.

A conceptual introduction to recommender systems

Businesses have a long-standing history of recommending their products or services to customers. For instance, walk into a bookshop and you are likely to see a list of popular books bought by other customers. This is a simple kind of recommendation system, as it gives buyers a snapshot of potential products to purchase.

In a bid to win in the digital economy, businesses are becoming increasingly customer-centric. **Customer centricity** implies that companies aim to put the needs of the customer first. Still, with the needs of customers being as diverse as the customers themselves, businesses need to take a unique approach in putting forward their products. This explains, in part, the failings of **popularity-based recommendation systems**, as they fail to consider the unique profiles of buyers. As such, with growing digitalization, increased business offerings, and a growing diversity of customers' needs, this approach is unlikely to win.

Interestingly, data science tools can offer a number of approaches to make recommender systems more intelligent by considering the needs of the buyers in a variety of ways.

In presenting the different types of recommender systems, we will continue to use the bookshop example.

First, the **item-based collaborative filtering** approach to recommendation systems makes product suggestions to book buyers based on the buyer's product purchase history and how those *products* relate to others. As such, if an individual bought *Book A*, and *Book A* is linked to *Book B*, then *Book B* is suggested. The second approach, **user-based collaborative filtering**, considers similarities between *buyers* when making suggestions. As such, if *Buyer A* is similar to *Buyer B*, and *Buyer A* buys *Book C*, then *Book C* would be recommended to *Buyer B*. The third approach, **content-based recommendation**, takes into account both the book and user characteristics in making suggestions. Finally, the **hybrid system** approach uses a combination of collaborative-based and content-based methods in making recommendations. It is easy to see that both of these methods come with strengths and weaknesses. We will now take a deeper look at these approaches and how DataRobot can be used to build a content-based recommendation system.

Approaches to building recommender systems

Recommender systems aim to suggest relevant products to buyers. Because of their ability to consider the uniqueness of buyers, intelligent recommender engines have generated billions of dollars for businesses and helped buyers find relevant products. They represent a win-win for both consumers and businesses. Various data-driven approaches to creating intelligent recommendation systems have been introduced. There are three major approaches to recommendation systems: collaborative filtering systems, content-based systems, and hybrid systems. Let's discuss each of these approaches in the following sub-sections.

Collaborative filtering recommender systems

The core idea behind collaborative filtering recommender systems is leveraging past actions by others to infer what an individual might be interested in. Collaborative filtering approaches draw on data stores of the historic interaction between products and users. *Table 10.1* presents an interaction matrix of users rating books. Each user rated a book with a number between 1 and 5, with 5 representing the highest level of enjoyment. Where there are no ratings, the individual is assumed not to have read the book. There are two broad types of collaborative filtering: *item-based* collaborative and *user-based* collaborative filtering.

	Book A	Book B	Book C	Book D	Book E
User 1	5		1		2
User 2	1	2	4		5
User 3	1			3	
User 4		1	5	4	
User 5					5

Table 10.1 – User/product interaction matrix

Item-based collaborative filtering systems (or **item-to-item collaborative algorithms**) find similarities between items and base their recommendations on these similarities. This approach is grounded in suggesting items to individuals based on how similar items are to the ones these individuals previously enjoyed or bought. Drawing on *Table 10.1*, an item-based filtering approach would easily see that *Book C* and *Book E* are rated in a similar way by previous readers. Based on this item relationship, if an individual rates *Book C* highly, a recommendation of *Book E* is made and vice versa. So, since *User 5* highly rated *Book E* and has not seen *Book C*, a recommendation of *Book C* is put forward, as there is a high likelihood of them liking *Book C*.

With *user-based collaborative filtering* systems, similarities are found between *users*, and recommendations are based on these. **User-to-user collaborative algorithms** aim to find users with similar behavior or who are in the same behavioral neighborhood, as established by their historic actions. The algorithm then considers what their preferences are and makes recommendations. The core idea of these recommendation systems is the assumption that if individuals *are* alike, *what* they like will be similar. From *Table 10.1*, it could be inferred that *User 2* and *User 4* have similar book interests. Because *User 4* has rated *Book D* highly, the likelihood of *User 2* liking *Book D* is considered high and therefore recommended. As we can see, both collaborative filtering approaches are based on the idea of *similarities*.

Similarity metrics offer a basis for recommendations to be made. There are several similarity metrics, with the **Pearson correlation coefficient** and **Cosine similarity** being among the most popular. Others have approached this measurement of similarity drawing on *neighborhoods*. The **K-nearest neighbors** algorithm is utilized to find the nearest items or users to the one being recommended or recommended to, respectively.

Because the interaction dataset is easily acquired, building collaborative filtering is considerably easier than content-based systems, as will be discussed in the next sections. However, the collaborative approach to recommendation systems has a few shortcomings. Within the context of *Table 10.1*, a new user, *User 6*, is introduced with no history. It is easy to see that the collaborative filtering system will struggle to make recommendations to this user. The problem is similar for an item without historic data. This problem, otherwise known as the **cold start** problem, is well documented. **Data sparsity** is another problem commonly associated with collaborative filtering. Most platforms and large businesses have buyers and products. Still, the most active users would only buy a fraction of the available products. As such, there is a gap in the data needed to meaningfully compute the similarities when powering these engines.

Content-based recommender systems

Content-based recommender systems make suggestions based on the item characteristics and user profiles. This approach has a different type of data structure underpinning it. Content-based systems are **machine learning** (**ML**) models, built by leveraging historic datasets consisting of item descriptions, user profiles, and user preferences. Some writers differentiate content-based recommender systems from *demographic* systems, but here, we consider demographic information as part of the profile of the user. In the case of buying, this classification model is used to predict the likelihood of users liking an item. Within the context of a books recommendation system, every book needs to be associated with its description, which could include its genre, cover, number of pages, size, and publisher, while information regarding the user could include their location, profession, age, and marital status. As illustrated in *Table 10.2*, the users' ratings come in addition to these. In this case, because the rating is represented by a value between 1 and 5, the model is regression-based, as it predicts an interval variable. This model becomes a content-based recommendation engine.

From the preceding discussion, we can see that a content-based system can easily mitigate the cold start problem, as books and users are likely to have some forms of descriptions. In comparison to collaborative filtering systems, content-based systems are more scalable, as in the production environment, predictions can easily be made when needed, rather than having to make predictions for all users and products at the same time. Importantly, even when users only rate or buy a few products, content-based systems will still perform well, as they focus on the descriptions and not necessarily the users or products. That said, most content-based systems struggle when the characteristics of the items are not readily available. Within certain contexts, it could be challenging to generate attributes for a product (for instance, if the product is or has images or sounds). In cases of this nature, content-based systems will have no descriptions to analyze. Additionally, demographic information of users might not be readily available due to growing online privacy concerns. The limitations of both the collaborative filtering and content-based approaches to recommendation gave rise to the use of hybrid systems.

Hybrid recommender systems

Hybrid recommender systems are an integrated approach to recommendation systems. Hybrid systems generate recommendations to users by leveraging a combination of two or more recommendation strategies. By doing so, they mitigate the limitations attributed to either of the strategies, thereby benefiting from *the wisdom of the many*.

There are several approaches to hybrid systems. The most commonly used (and the easiest to implement) is the **weighted approach**. Here, scores from independent recommendation systems are aggregated to give an overall recommendation score. Aggregation methods vary and can include basic averaging, applying rules, and using linear functions. The **staged approach** could also be deployed. This typically involves the recommendation systems' results being integrated as input features in another recommendation system. As such, the output of the *Stage 1* system becomes an additional input for the *Stage 2* system. The **switching approach** involves using a rule to switch between different recommendation systems to capitalize on their advantages in a given context. For instance, if collaborative filtering is seen to give better results, a switch regime could use the collaborative filtering approach, but when there is a cold start, it could change to the content-based approach. An advantage the hybrid system has over content-based systems is the ability to develop recommendations when item features are difficult to establish. As will be demonstrated in *Chapter 11*, *Working with Geospatial Data, NLP, and Image Processing*, DataRobot has advanced feature extraction capabilities for images and text data.

Defining and setting up recommender systems in DataRobot

DataRobot, due to its ability to extract features from images, audio, and text data, effectively manages the feature availability limitation of the content-based recommender systems. This, in addition to DataRobot's automated ML models' processes, means it is well positioned to leverage the advantages of the content-based approach while compensating for the feature-unavailability limitation of this approach. As described in the *Technical requirements* section, the dataset used for our example consists of three tables. This includes the user table (presenting profiles of the users), the book table (outlining characteristics of the books), and the rating table (containing user book ratings). Since we have one table describing the books, and another, the users, integrating these and the ratings sets the scene for the content-based recommender system. To do this, we employed Jupyter Notebook. *Figure 10.1* presents the script we ran to ingest the dataset, manipulate it, merge the tables, and write it back as a CSV file:

```python
import numpy as np
import pandas as pd
books = pd.read_csv("Books1.csv", sep=',', encoding="latin-1", error_bad_lines=False)
users = pd.read_csv("Users.csv", sep=',', encoding="latin-1", error_bad_lines=False)
ratings = pd.read_csv("Ratings.csv", sep=',', encoding="latin-1", error_bad_lines=False)
```

```python
books = books[['ISBN', 'Book-Title', 'Book-Author', 'Year-Of-Publication', 'Publisher']]
books.rename(columns = {'Book-Title':'title', 'Book-Author':'author', 'Year-Of-Publication':'year', 'Publisher':'publisher'},
        inplace=True)
users.rename(columns = {'User-ID':'user_id', 'Location':'location', 'Age':'age'}, inplace=True)
ratings.rename(columns = {'User-ID':'user_id', 'Book-Rating':'rating'}, inplace=True)
```

```python
ratings = ratings.merge(books, on='ISBN').merge(users, on='user_id')
```

```python
ratings = ratings[['user_id', 'ISBN','title', 'author', 'year', 'publisher',
        'location', 'age', 'rating']]
```

```python
ratings.head()
```

```python
ratings.to_csv("ratings.csv")
```

Figure 10.1 – Data manipulations in Jupyter Notebook

Rows on the rating table where `rating` had a value of `0` were excluded, as the ratings were implicit. These rows will be used to demonstrate how to make predictions with recommendation engines in the *Making recommender system predictions with DataRobot* section. Having manipulated the tables by changing their headings, as well as consolidating the `ratings`, `books`, and `users` values into a table, each row has the description of a user and a book, and also a rating. A snapshot of the data is shown in *Table 10.2*. Although we could create the DataRobot project in Jupyter Notebook using the Python API method (as will be illustrated in *Chapter 12, DataRobot Python API*) for consistency, we downloaded the data as a file: `rating.csv`.

	user_id	ISBN	title	author	year	publisher	location	age	rating
0	2313	034545104X	Flesh Tones: A Novel	M. J. Rose	2002	Ballantine Books	cincinnati ohio usa	NaN	5.0
1	2313	812533550	Ender's Game (Ender Wiggins Saga (Paperback))	Orson Scott Card	1986	Tor Books	cincinnati ohio usa	NaN	9.0
2	2313	679745580	In Cold Blood (Vintage International)	TRUMAN CAPOTE	1994	Vintage	cincinnati ohio usa	NaN	8.0
3	2313	60173289	Divine Secrets of the Ya-Ya Sisterhood : A Novel	Rebecca Wells	1996	HarperCollins	cincinnati ohio usa	NaN	9.0
4	2313	385482388	The Mistress of Spices	Chitra Banerjee Divakaruni	1998	Anchor Books/Doubleday	cincinnati ohio usa	NaN	5.0
5	2313	399146431	The Bonesetter's Daughter	Amy Tan	2001	Putnam Publishing Group	cincinnati ohio usa	NaN	5.0
6	2313	345348036	The Princess Bride: S Morgenstern's Classic Ta...	WILLIAM GOLDMAN	1987	Del Rey	cincinnati ohio usa	NaN	9.0
7	2313	553278223	The Martian Chronicles	RAY BRADBURY	1984	Spectra	cincinnati ohio usa	NaN	7.0
8	2313	449912558	The Sparrow	MARY DORIA RUSSELL	1997	Fawcett Books	cincinnati ohio usa	NaN	0.0
9	2313	20442602	Voyage of the Dawn Treader	C. S. Lewis	1970	MacMillan Publishing Company.	cincinnati ohio usa	NaN	9.0

Table 10.2 – Data snapshot

Following the process established in *Chapter 6, Model Building with DataRobot*, we created a DataRobot project for the recommender system. When doing this, we drag the `rating.csv` file into the initial project window. This opens up the window shown in *Figure 10.2*. For each row, since the book rating is used as an indicator of the user's interest, it can be used as the target variable. Due to the nature of the target variable, `ratings`, the ML models for this recommender system will be of the regression models type.

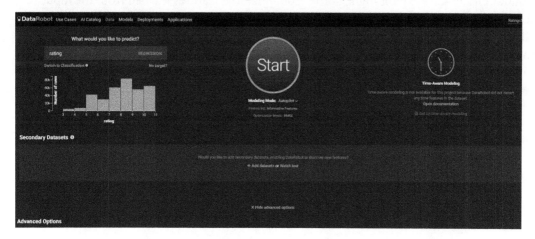

Figure 10.2 – DataRobot project initiation window

As expected, ratings made are the range of 1 to 10. Ideally, we will drop the rows with implicit ratings (of 0) and user_ID fields to create a robust dataset for modeling. The next thing to do is build the recommender system's ML models.

Building recommender systems in DataRobot

One of the strengths of driverless **artificial intelligence** (**AI**) platforms such as DataRobot lies in their simplification of the data science model-building process. Given the similarity of the content-based recommendation model-building process to the typical ML one, DataRobot's ML capabilities could be leveraged in building these systems. Having set up the data (as detailed in the previous section), click on **Start** (*Figure 10.2*) to commence the modeling process. To avoid over-optimistic model performance which fails to generalize where users provide more than one ratings for items, it might be useful to partition the rating according to the users. To do this, within the **Advanced options** window, open the **Group** tab and enter user_id in the **Group ID Feature** field.

As detailed in *Chapter 6*, *Model Building with DataRobot*, DataRobot commences the development of ML models when the **Start** button is clicked. However, with recommender systems, the DataRobot's present strong prediction accuracy as the platform benefits from cutting-edge technological advances in developing models. Recommendation datasets pose difficulties for modeling because of their high data sparsity and dimensions. The DataRobot's models exploit higher-order combinatorial features learned from the input data. Though some of these models will not run automatically when the **Start** button is clicked, they can be accessed in the **Repository** tab. Because these models are based on the **Keras** neural network, they use a training schedule in their development. So, they can easily be found by entering Training Schedule or simply Training in the search field in the model **Repository** tab during model creation. This will bring up a list of relevant models (see *Figure 10.3*):

Figure 10.3 – Selecting the advanced modeling approaches most suitable for recommender systems

In addition to selecting these modeling methods to be included in the list of models to be created, the models' **Sample Size**, **Cross Validation runs**, and **Feature List** options are to be set For the current project, we selected 16% of the sample size (which snaps to `Autopilot Stage 1`) based on the `Informative Features`, and then carried out all five `Cross Validation` runs. A final click on **Run Tasks** includes these in the processing queue.

After the models have been created, the next step is to evaluate them in terms of their accuracy. Prior to this, it is important to examine the **Relative Importance** chart to check if our model aligns with common sense. As is apparent in *Figure 10.4*, opening the **Variable Importance** window through the **Insight** window offers us the opportunity to explore these models:

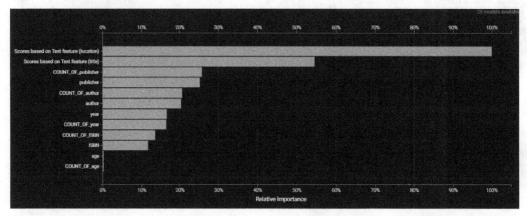

Figure 10.4 – Variable importance

The values next to **Scores based on Text feature (location)** and **Scores based on Text feature (title)** suggest that the model performance is significantly informed by a person field and an item field. Models that capitalize on the learned creation of higher-order variables excel in these situations. This is because they generate higher-order variables that are interactions between the person-specific and item-specific features, drawing on learning from the data. As a result, in the preceding case, a feature that combines the presence of an aspect of a location (for example, London) and an aspect of the title (for example, Kingdom) could be interpreted as influential to the model. So, in this simplified example, a higher-order feature that is an interaction between London and Kingdom is created. The rating predictions consequently change considerably depending on the presence of this newly created higher-order feature.

In model selection, using the definite **root mean squared error** (RMSE) evaluation metrics, we see that `Keras Slim Residual Neural Network Regressor using Adaptive Training Schedule (1 Layer: 64 Units)` is the best-performing model (see *Figure 10.5*). It is important to highlight that measuring the accuracy of models for recommendation systems in some contexts is not as straightforward. Imagine that in this case, we could only have a rating of 1 when an individual buys a book, and otherwise it would be 0. Naively measuring how accurate the model is becomes limited, as a 0 rating does not necessarily imply that an individual is not interested in an item. This is because it is possible that the individual has never read the book. Because a good recommender system will recommend items whose characteristics align with an individual's profile as a potential book to read that are unread, it is likely to have a significant proportion of false positives. This is because, although their current rating is 0, the user in question will most likely be interested in reading them. In cases like these, the **Recall** type becomes a more important metric in evaluating the model performance. Given that we are only certain of cases where individuals buy items, it is reasonable to evaluate those cases in isolation. Therefore, the extent to which the model accuracy predicts books that are read correctly, usually referred to as the **Recall**, becomes a more suitable metric.

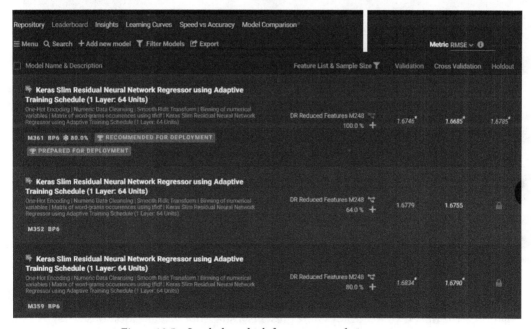

Figure 10.5 – Leaderboard tab for recommendation systems

For a recommendation system, accuracy and prediction speed is very important to consider when deciding which model to use. To ground this discussion, it is important to understand that there are two major approaches to making predictions with recommendation systems. The first approach is a batch scoring of combinations of users by items, where items are yet to be read by the user. This dataset becomes larger exponentially as items and users increase. The second approach is a *real-time prediction*. For instance, imagine an individual arrives at an e-commerce platform. That individual's data with those of the products is rapidly scored and suggestions are scored nearly instantly. In both cases, the speed of the prediction is pivotal for commercial success. The DataRobot **Speed vs Accuracy** chart offers some support in analyzing speed and accuracy for recommendation systems. As seen in *Figure 10.6*, the RMSE metric for that `Keras Slim Residual Neural Network Regressor using Adaptive Training Schedule (1 Layer: 64 Units)` is `1.6746`, and its prediction speed is `35.57` ms per every 1,000 predictions. The validation scores for some blender models appear better, but these are much weaker in terms of the speed of prediction.

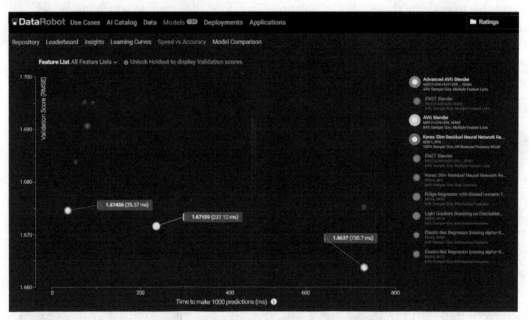

Figure 10.6 – DataRobots' Speed vs Accuracy chart

This suggests that though it is very accurate, this model is very slow in making predictions. The **Speed vs Accuracy** chart presents a snapshot visualization of several models' speed and accuracy. A more in-depth pairwise comparison can be carried out using the **Model Comparison** tool. To continue the discussion of prediction, we will now turn to making recommendation system predictions in DataRobot.

Making recommender system predictions with DataRobot

Creating suggestions from recommendation engines on DataRobot is straightforward. We use the drag and drop approach (as discussed in earlier chapters), as our prediction dataset is only small. With larger datasets (over 1 GB), as is more typical for recommender systems, using the DataRobot prediction API is advised. The API approach to creating models and making predictions is covered in depth in *Chapter 12, DataRobot Python API*.

Our prediction dataset for our example is 64 MB in size, and so the drag and drop approach is appropriate. For this prediction approach, we specify the columns we want to use from the original dataset. Ideally, we at least need an identifier for the item and user. As illustrated in *Figure 10.7*, we have chosen to include the ISBN, user_id, and title fields in our predictions. We drag and drop the prediction dataset into the specified region. As usual, this dataset is quickly evaluated, and we are presented with the **Run external test** or **Compute prediction** options.

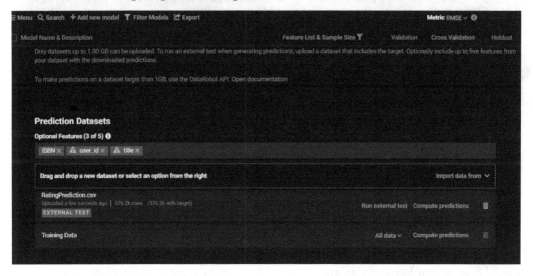

Figure 10.7 – A recommendation engine prediction setup

At this point, we click on **Compute predictions** to commence the prediction process. Following the completion of the prediction process, the predictions are downloaded as a `.csv` file (see *Table 10.3*). As noted previously, the prediction set is drawn from the original dataset where the ratings were implicit (so the rating score was zero). Thus, the prediction dataset has only a limited sample of the possible person-item interactions. Some users (for instance, the user with `8` as `user_id`), have about `10` items scored, while some have only `1` item scored. In an ideal situation, all items not seen by an individual would be rated. That said, suggestions served to the user are then made in order of predicted interests. For user `8`, the book titled `A Second Chicken Soup for the Woman's Soul (Chicken Soup for the Soul Series)` is served first. In some cases, the top *n* recommendations is used. By top *n* in our book case, we mean, for each user the top *n* books are selected based on their prediction values.

	A	B	C	D	E
1	row_id	ISBN	user_id	title	Prediction
2	5E+05	195153448	2	Classical Mythology	7.967735
3	5E+05	1558746218	8	A Second Chicken Soup for the Woman's Soul (Chicken Soup for the Soul Series)	6.707153
4	5E+05	679425608	8	Under the Black Flag: The Romance and the Reality of Life Among the Pirates	6.354887
5	5E+05	080652121X	8	Hitler's Secret Bankers: The Myth of Swiss Neutrality During the Holocaust	6.326581
6	5E+05	393045218	8	The Mummies of Urumchi	6.252465
7	5E+05	771074670	8	Nights Below Station Street	6.090163
8	5E+05	399135782	8	The Kitchen God's Wife	6.045892
9	5E+05	671870432	8	PLEADING GUILTY	5.944796
10	5E+05	425176428	8	What If?: The World's Foremost Military Historians Imagine What Might Have	5.839644
11	5E+05	60973129	8	Decision in Normandy	5.809216
12	5E+05	374157065	8	Flu: The Story of the Great Influenza Pandemic of 1918 and the Search for the V	5.437757
13	5E+05	440234743	9	The Testament	7.788191
14	5E+05	609804618	9	Our Dumb Century: The Onion Presents 100 Years of Headlines from America's	7.426695
15	5E+05	1841721522	10	New Vegetarian: Bold and Beautiful Recipes for Every Occasion	7.98113
16	4E+05	971880107	14	Wild Animus	4.005665
17	5E+05	345417623	16	Timeline	6.918756
18	5E+05	553278398	17	Prelude to Foundation (Foundation Novels (Paperback))	7.516078
19	5E+05	684823802	17	OUT OF THE SILENT PLANET	7.013817
20	5E+05	312978383	17	Winter Solstice	7.012969
21	5E+05	425163091	20	Chocolate Jesus	6.838427
22	5E+05	3442353866	22	Der Fluch der Kaiserin. Ein Richter- Di- Roman.	6.672094

Table 10.3 – A recommendation engine sample prediction

The selected model can be deployed as a **REST** API using DataRobot, as shown in *Chapter 8, Model Scoring and Deployment*, and then the data can be scored via the DataRobot API call (which we will discuss in *Chapter 12, DataRobot Python API*). Some DataRobot models can be downloaded as **JAR** files, which can be integrated with other applications to make real-time predictions. Elsewhere, a batch prediction can be made using different person-item interactions, before being stored in a big data storage table, such as **Google Cloud BigQuery**.

Summary

In this chapter, we introduced and appraised different approaches to recommendation systems. We examined the data structure requirements for content-based and collaborative filtering recommendation systems, and we discussed their underlining assumptions. We then point out the strengths of DataRobot in extracting features from challenging data types (for instance, image data) that normally limit the use of content-based systems. We then illustrated the use of DataRobot in building and making predictions using a content-based recommender system based on a small dataset.

It is important to highlight that the dataset used for this project was made up of multiple data types. DataRobot is capable of extracting features and integrating different data types to create ML models. In the next chapter, we will explore how to use datasets with a combination of image, text, and location data when creating ML models.

11
Working with Geospatial Data, NLP, and Image Processing

In this book thus far, we have focused mainly on numeric and categorical features. This is not always the case in big data, as with big data comes an increasing data variety. Image, text, and geospatial data is becoming increasingly valuable in gaining insight and providing solutions to the most complex problems. Recently, for instance, **location-based** data has been used to improve the effectiveness of advertising campaigns. For example, different ads can be shown to users according to their location; if they are coffee lovers and close to coffee shops, push notifications could be sent to their mobile devices. In other cases, chatbots, based on advanced text analytics or natural language processing, provide businesses with an efficient and effective avenue to solve customer problems. What is most interesting and an emerging approach to solving commercial problems is the use of **multimodal** datasets, which combine different variable types in the same project.

Understandably, the topic of analyzing different variable types is enough to be covered in a book in its own right. Yet providing an overview of the analysis of different variable types is key in grounding the use of DataRobot in building multimodal models that involve text, image, and location data. With that foremost in mind, in this chapter, we will delve into the definitions and approaches to analytics with text, image, and geospatial data. Thereafter, we will use DataRobot to build and make predictions with a model that capitalizes on the uniqueness of a multimodal dataset in predicting house prices. As such, the topics that will be covered are as follows:

- A conceptual introduction to geospatial, text, and image data
- Defining and setting up multimodal data in DataRobot
- Building models using a multimodal dataset in DataRobot
- Making predictions using multimodal datasets in DataRobot

Technical requirements

Most of the analysis and modeling carried out in this chapter requires access to the DataRobot software. Some manipulations were carried out using other tools, including MS Excel. The dataset utilized in this chapter is the House Dataset.

House Dataset

The House Dataset can be accessed at Eman Hamed Ahmed's GitHub account (`https://github.com/emanhamed`). Each row in this dataset represents a specific house. The initial feature set describes its characteristics, price, zip code, images of the bedroom, bathroom, kitchen, and frontal view. There was no missing data. We went on to develop text descriptions for each house, based on the number of bedrooms, bathrooms, city, country, state, and actual size of the property. Elsewhere, the ZIP codes were converted into latitude and longitude, which were added to the dataset as columns. More information on the base features is provided at the GitHub link and the data is provided in `.csv` format.

Dataset Citation

House Price Estimation from Visual and Textual Features. In Proceedings of the 8th International Joint Conference on Computational Intelligence, H. Ahmed E. and Moustafa M. (2016). (IJCCI 2016) ISBN 978-989-758-201-1, pages 62–68. DOI: 10.5220/0006040700620068

A conceptual introduction to geospatial, text, and image data

Just like we use different senses to holistically understand objects around us, a **machine learning** (**ML**) model also benefits from data coming from different types of sensors and sources. Having only one type of data (for instance, numeric or categorical) limits the level of understanding, predictability, and robustness of a model. In this section, we will present a more in-depth discussion of the business importance of different data types in building models, the associated challenges, and the preprocessing steps necessary to mitigate these challenges.

Geospatial AI

Geospatial understanding has had long-standing implications for decision-making in certain industries, including mineral exploitation, insurance, retail, and real estate. While the commercial importance of data science is well established, location-based AI is just beginning to gain recognition. The use of ML in improving business performance has brought to the fore the importance of augmenting datasets with location-based information and features in building predictive models.

Typical ML models built mainly from categorical and numeric data have contributed immensely to realizing business goals, but decisions are governed by more than numeric and categorical information. Indeed, the events take place at certain locations. ML models need location-based information in order for the location context to effectively present commercial insight and predictions. What works in one geography may not work in another.

The potential commercial impact of using ML and location-based information comes with several challenges:

- A lack of datasets, tools, and people skills.
- Connecting ML pipelines to native location-based analysis techniques is not straightforward.
- Only a few R and Python packages have geospatial capabilities.
- Understanding these capabilities requires further education and training for analysts.

DataRobot's location AI capability helps alleviate some of these challenges. The location AI capability complements the existing AutoML experience by adding in a repertoire of geospatial analytic and modeling tools. With DataRobot, location features could be selected from the dataset, but the location AI capability enables the platform to automatically recognize geospatial data and create geospatial features. A variety of geospatial data file formats can be uploaded. These include GeoJSON, Esri shapefiles, and geodatabases, PostGIS tables, as well as traditional latitude and longitude data.

Natural language processing

As humans, we communicate effectively via a vast range of words with or without limitations on the volume of words to use. More than words, body language, tonality, and words' context are crucial to effective communication. For example, using the same set of words, *the cat is bigger than the dog* has a different meaning to *the dog is bigger than the cat*. Naturally, humans understand, draw conclusions, and make predictions of the future based on free text. The use of free text comes with valuable information and rich insights can be harvested from it. Yet, since free text fails to follow a consistent structure, they pose challenges to being processed by machines.

Conversations and other forms of free text are messy and unstructured as they do not fit neatly into traditional tables with rows and columns. **Natural Language Processing (NLP)** sits at the intersection of data science and linguistics and involves the systematic use of advanced processes for analysis, understanding, and the extraction of data from free text. Through NLP, scientists can leverage free text to generate valuable insight, which is then integrated as features in building better-performing models. Text mining allows the identification of unique words or groups of words that are associated with certain outcomes. For example, in the house price prediction case, the description of the house improves the predictability of the models in estimating the house price. Thinking about it, the description also contributes to an individuals' decision of buying a house. Individuals' propensity to buy houses influences property pricing. NLP algorithms can identify the effect of word sequencing and influence words or phrases, and a word's context within sentences.

NLP is key to machines being able to extract important information from text. Consequently, NLP allows machines to decide feelings described in free text by giving a number score to a text, indicating its sentiment to a topic or event. Similarly, it aids in the identification of classes that certain words most likely belong to. This capability has given birth to several applications, including text classification, named entity recognition, sentiment analysis, and summarization of text.

To get free text to provide useful insight or be integrated into models is not an easy task. As earlier alluded to, raw text has no structure, so structure needs to be introduced. Also, numerous words have the same meaning and you could have the same word mean different things in different contexts. In a typical analytics process, there are numerous steps taken to normalize free text. At least four steps are required:

1. The first step in most text processing is splitting text corpus into separate words. This step, also called **tokenization**, enables the identification of keywords and phrases. The separated words are referred to as tokens. N-grams are the basic units for text analytics.

2. Next, there are certain words that contribute little or nothing to the meaning of a text. These are generally common words; for instance, in the English language, we have words such as *the, that, is,* and *these.* Within the context of text mining, these words are referred to as noise or are sometimes called stop words. So, this step is called **noise removal.**

3. After that, words are converted to their root meanings. There are a few approaches to this. As an illustration, **stemming** typically converts to the root word stem by eliminating certain letters. So, words like *happy, happiness, happily,* and *happiest* will all be returned to the root word *happ.* Because the same words could have multiple meanings, disambiguation of words becomes crucial in text processing. Whereas stemming returns words to their roots by cutting off their prefix or suffix, **lemmatization** examines the context of words to ensure stemmed words are converted to logical bases called **lemma.** For example, the word *anticipate* when stemmed might be returned to *ant.* Within the context, however, *ant* would not make sense; as such, lemming will ensure that the word *anticipate* is retained.

4. A final, yet important, step is the process of **featurization** where lemma or root words are converted into features. Again, there are several ways this can be done. The most straightforward method involves developing features for each unique token and counting the number of that token in each text corpus (*Table 11.1* presents a demonstration of this process):

	Blue	Black	Green	Car	Truck
Blue Car	1	0	0	1	0
Black Car	0	1	0	1	0
Blue Truck	1	0	0	0	1
Blue Black Car	1	1	0	1	0
Green Truck	0	0	1	0	1

Table 11.1 – A demonstration of featurization

Following featurization, developed text variables are either used as predictors alone or integrated with other variables in building models. While the importance of analytics with free text within the commercial setting is well established, ancient wisdom suggests that *"a picture is worth a thousand words."* This raises the importance of using image analytics in driving business value. As we have established in this section, the core purpose of NLP is the extraction of text features from raw text. Image processing performs a similar function for images, as described in the next section.

Image processing

Images provide valuable information to customers about products and services. Images are fast becoming crucial to business success as they influence the propensity to buy. As the data landscape continues to grow, image data is also becoming readily available and important. This offers analysts the opportunity to include image features when creating insights for businesses.

Image data, like text data, lacks structure. In fact, with image data, there is an uncertainty of features. To bring this to life, let's imagine the case of identifying individuals from their pictures. The image of an individual could be colored or grayscale, the position of an individual's face or body could change, and their background and outfits are unlikely to be the same across images. These and other variations make the data generated from differing images of the same person appear very different. As such, features from an image are unlikely to be consistent with those of another image, despite the individual being the same person; therefore, the image data would have challenges with feature uncertainty. Yet, the human eye can see the individual in those images as the same and easily recognize them.

The smallest indivisible units within images are known as **pixels**. For grayscale images, pixels are interpreted as 2D arrays. Each has a strength represented by a value between 0 and 255, referred to as **pixel intensity**. For grayscale images, 0 is shown as completely black, while completely white gives 255. On the other hand, color images have 3D arrays with blue, green, and red layers. Like black on the grayscale images, each of those layers has its own values from 0 to 255, where the final color is a combination of corresponding values on each of the three layers.

Image processing typically follows predefined steps in extracting useful and consistent features from images that align with the purpose of extraction. **Activation maps** are then applied to the extracted images to reduce the computational load required to process the volume of data from each image. As such, the activation maps essentially reduce the feature space. The reduced feature space is then **flatted** into a tabular structure, enabling it to be used as variables in the typical ML modeling process. Though this is a simplified illustration of image processing, there are other approaches to it. This ensures we have some understanding of how the challenge of image inconsistency and limited structure can be managed.

We have so far established the added value of geospatial, text, and image data. We have discussed the challenges in using these data types and also highlighted the key steps in using them to build models. Sometimes, different data types such as images and text are integrated into the same dataset in training models. This type of dataset is referred to as a **multimodal dataset**. While a multimodal dataset presents unparalleled opportunities, it comes with huge challenges. The difficulties and steps highlighted for each of the data types are expected to be addressed. DataRobot has capabilities that make use of multimodal data more accessible. The platform enables preprocessing steps and integrates these datasets in making predictions. For the rest of the chapter, we will demonstrate how to use the price listing dataset to train a model and make predictions.

Defining and setting up multimodal data in DataRobot

DataRobot's location AI, text mining, and visual AI automated ML capabilities make working with text, location, and image data relatively straightforward. With these capabilities, DataRobot can help analysts in building models and making predictions against text, image, and location-based datasets. Data set up for an image model differs considerably from other types of models. Using our House Dataset, our first task is to set up the data.

The house dataset has zip codes for the houses. We created and integrated latitude and longitude coordinates from the zip codes as new columns. Other features created from the zip codes not evident in *Table 11.2* are the city, county, and state where each of the houses was located. Further text description columns were built from the number of rooms, the size of the house, and its location:

Table 11.2 – The developed price list dataset

The original data comes with 2,140 images, each of the 535 houses having a bedroom, bathroom, frontal view, and kitchen. For the image data to be included in the analysis, the data can be set up as a ZIP file with a structured .csv file sitting next to the image folder. Four new image columns were created for the bedroom, bathroom, frontal view, and kitchen. As such, each data row on the .csv file had paths to their images, as shown in *Table 11.2*. The paths point to locations on the corresponding image file in the ZIP file. Within the ZIP file, the .csv file with the tabular features sits needs to the folder, containing all the images. Each image has a unique name that is consistent with the image path columns on the .csv file. The setup for the ZIP file and HousePrice folder is shown in *Figure 11.1*. That said, the dataset could still be ingested using the AI Catalog. This also gives DataRobot the ability to connect to other data sources for the images. Furthermore, the Paxata tool can be deployed in the data preprocessing if you have access to that tool:

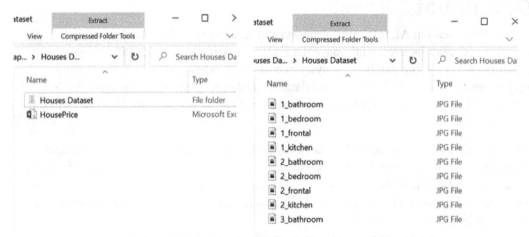

Figure 11.3 – Data setup for the ZIP file (left) and image folder (right)

The image on the left of *Figure 11.1* shows how the ZIP file is set up. The folder containing the image files, Houses Dataset, is next to the HousePrice.csv file. The image on the right presents image files within the Houses Dataset folder. Here, the images are labeled, with the locations, consistent with cells on data, HousePrice.csv (as shown in *Table 11.2*). With the data completely set up, the next step is to commence model development.

Building models using multimodal datasets in DataRobot

Having fully set up our ZIP file with the multimodal dataset, we proceed into initiating the project within DataRobot. The data ingestion using the drag and drop method is like the earlier project, except in this case we upload the ZIP file. Following the upload of the ZIP file, the price is selected as the target variable. DataRobot automatically detects the text, image, and geospatial fields (see *Figure 11.2*). The geometry feature is a location-based feature made up of the latitude and longitude variables in the original dataset. Apart from latitude and longitude coordinates, location features can be formed from other native geospatial formats, such as Esri shapefiles, GeoJSON, and PostGIS databases. These can be uploaded using drag and drop, AI Catalog, or URL methods:

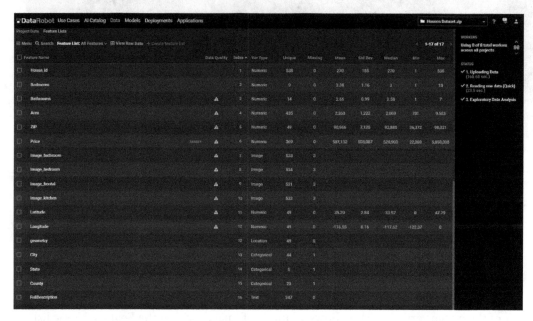

Figure 11.4 – Feature Name list

5. The location-based visual representation of the listing price can be viewed by
    selecting the **Price** option in the **Feature Name** list. This **Exploratory Spatial Data
    Analysis** (**ESDA**) is conducted by opening the **Geospatial Map** tab and clicking on
    the **Compute feature over map** button. As seen in *Figure 11.3*, the **Geospatial Maps**
    window offers a location-based analytics visualization of the dataset. It shows the
    distribution of properties over space – the number of houses in each area and their
    average prices:

Figure 11.5 – Geospatial Map

The map legend offers vital information about the map. It highlights that the color
of the hexagon shows the average house prices within the location. Elsewhere, it
presents the frequency of cases by the height of the hexagons. This ESDA feature
shows not only the visual distribution of house prices across the map but also an
illustration of house counts in differing areas. Similar geospatial analysis can be
conducted for other features, such as house area variables and bedrooms.

This preliminary examination of image features can be conducted by selecting any
of the image variables. This shows a sample of images within the **Feature Name** list.
Here, differing image features can be seen and organized by house price ranges.

6. To further explore the image features at a property level, we click on **View Raw Data**. This opens the dataset in its final format on DataRobot. Unlike the initial .csv file with image paths, the images are integrated into the dataset (see *Figure 11.4*). For each of the rows, the images are clearly displayed. A further scroll will show the free text description of the listed properties. This multimodal dataset of text, location, and image features can now be used to build a more robust model and make predictions of house prices:

Figure 11.6 – The DataRobot view of the multimodal data

7. As with earlier projects, we click on **Start** to commence the model-building process. On completion of the modeling process, the models are evaluated using the **RMSE metric**. The leaderboard shows DataRobot has built 36 models in total. The top-performing is the **Nystroem Kernel SVM Regressor** model.

As can be seen in *Figure 11.5*, opening the model presents its blueprint, outlining all the steps necessary to make the data ready for this model. Because of the multimodal nature of the data, the preprocessing steps are quite complex. DataRobot conducted geospatial processing, which was integrated with some numeric variables and high-level image and text processing (the latter not visible in *Figure 11.5*). For more information on each step, a click on the step box provides some insight on the modeling step and a link to comprehensive documentation on the step:

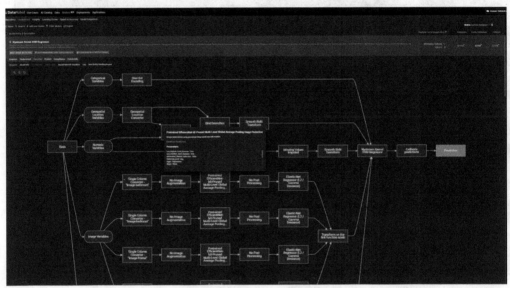

Figure 11.7 – The model blueprint for multimodal data modeling

8. Within the **Understand** tab, the **Feature Impact** view highlights the extent to which features contribute to the overall performance of the model (see *Figure 11.6*). The **Feature Impact** view for this project shows that `Area` is the most impactful feature of the house; next is the `FullDescription` text feature. Thereafter, the `Bedrooms` and `Image_kitchen` features follow suit. What is rather interesting is the fact that `Image_bathroom` seems to have a negative impact on the model accuracy. This suggests that insights from these images lead the model away from actual house prices:

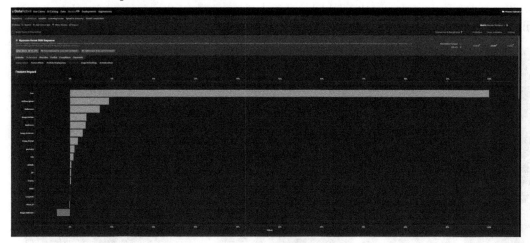

Figure 11.8 – Feature Impact for multimodal models

For this reason, leading the model away from improved performance, we use the **image embedding** and activation maps to understand how the model uses `bathroom` images to make predictions. By doing so, we will use the `bathroom` feature to demonstrate the image feature exploration capabilities available in DataRobot. DataRobot conducts unsupervised learning to cluster images according to their similarity. Still within the **Understand** tab, this is presented within the **Image Embeddings** sub-tab for each image feature. *Figure 11.7* presents the image embedding for the `bathroom` views. We can see DataRobot clusters similar images together. It seems that images that are dominantly white goods are presented in the right-hand and upper parts of the visualization. We can filter the visualization in accordance with house prices:

Figure 11.9 – Image embedding

Activation Maps adds to this information by offering insight into what aspect of images the model is leveraging in making predictions. This is critical to confirm that the model is using the key aspects of images. *Figure 11.8* presents the activation map for the `image_bathroom` variable. It appears the model makes its predictions mainly from white fixtures in the bathroom. This might offer insight into why `image_bathroom` is seen as having a negative impact on the model performance. It is possible that this extraction from white fixtures misleads the model:

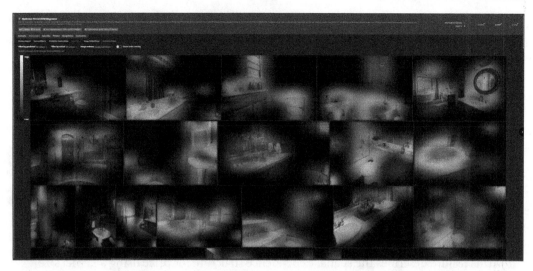

Figure 11.10 – Activation map

Location-based information comes with significant information complementing other data types. However, some models struggle in certain geographical areas. Inspecting the performance of models and considering locations empowers the analyst in taking actions on model performance improvement. DataRobot's **Accuracy Over Space** capability presents a spatial representation of a models' residual at differing locations (see *Figure 11.9* for an example). This chart could lead the analyst into considering the rationale behind higher residuals in certain areas:

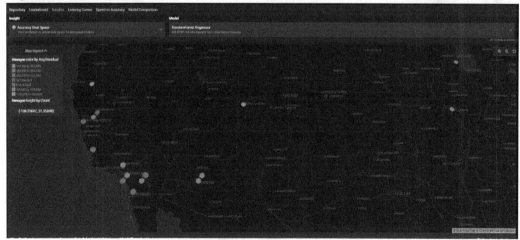

Figure 11.11 – Accuracy Over Space

For instance, as evident in *Figure 11.9*, the Phoenix area has a higher average residual price of over \$380k than most places. This area might for instance be considered an area of low income. This visualization could point the data scientist toward including features around localized economic indices. This might provide an explanation for the high residual. Including such features could therefore improve the overall performance of the model. The data partition for measuring accuracy could be set by altering between the validation, cross-validation, or holdout partitions. Also, the accuracy metric type and aggregation can be adjusted in accordance with the user's requirements.

Complementing its location-based features engineering, DataRobot's location-based analytics capabilities can exploit its location awareness to create **spatially autocorrelation features**, sometimes known as **spatially lagged features**, which are extremely insightful. The `eXtreme Gradient Boosted Trees Regressor (Gamma Loss)` model's fourth most important feature, `GEO_KNL_K10_LAG1_Price`, is one such feature (see *Figure 11.10*). This feature describes the spatial dependence structure for price using a kernel size augmented by distance. The **k-nearest neighbor** approach can also be deployed:

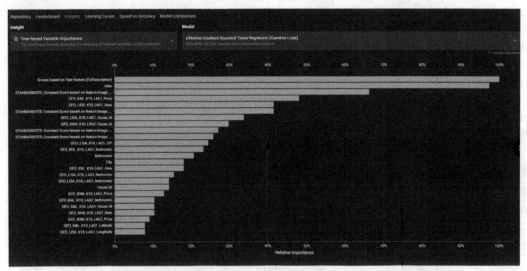

Figure 11.12 – Spatial lag features

Text analytics information such as the Word Cloud is not available within the **Understand** tab for this model. We turn to the **Insights** view to learn more about the `FullDescription` text feature, which is indeed one of the most impactful features of this model. Though not visible on the model blueprint (*Figure 11.10*), the text variable was scored using another model, `Auto-Tuned Word N-Gram Text Modeler using token occurrences - FullDescription`, which essentially develops scores using **N-Gram** and their token occurrence. During this modeled step, the `FullDescription` feature was converted into tokens for a differing number of words (as *N* in *N-grams*) and scored. Thereafter, this feature was transformed on the link scale and standardized. For text-related insights, we turn to the **Insights** view, offering two important text insight capabilities, **Word Cloud** and **text mining**.

The **Word Cloud** provides a diagrammed representation of the effect of certain words or groups of words (otherwise referred to as tokens) within the `FullDescription` feature in influencing the house price. The size of the words, as shown in *Figure 11.11*, highlights the frequency of the tokens, while the color suggests its effect coefficient. This coefficient is standardized typically between -1.5 and 1.5. The closer the color of the words is to red, the greater the coefficient and, consequently, the greater the house price is:

Figure 11.13 – Word Cloud for a multimodal dataset

We can assume that when the `FullDescription` variable contains words in orange, **alameda county**, and big, the prices are likely to be high. Similarly, with small-sized words, and **city riverside**, a lower price is expected. The text mining capability displays similar information to the Word Cloud using a bar graph.

Now that we have been able to build models using multimodal datasets, conduct analysis on their features, and evaluate the performance of those models, we will next focus on making predictions with models.

Making predictions using a multimodal dataset on DataRobot

After building a model, there are many ways to make predictions on a DataRobot. For this use case, we will illustrate the prediction capability using the `Make Prediction` method, which is available within the **Predict** tab. We initially create a prediction ZIP file dataset using the step outline in the *Defining and setting up multimodal data in DataRobot* section of this chapter. The developed prediction dataset is either dragged and dropped into the highlighted area or locally imported. As seen in *Figure 11.12*, we select the features we are interested in, including the prediction dataset:

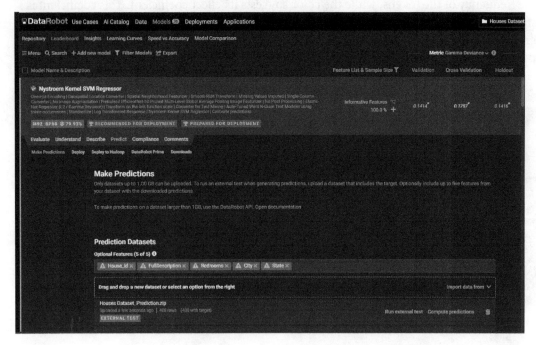

Figure 11.14 – Making a prediction from multimodal datasets

In this illustration, we selected `House_id`, `FullDescription`, `Bedrooms`, `City`, and `State`. We can also see that the prediction dataset has 400 houses. Finally, **Compute predictions** is selected to make predictions. When predictions have been completed, they are downloaded. This straightforwardly creates a downloadable `.csv` file, which has all the requested columns (see *Table 11.3*):

	A	B	C	D	E	F
1	House_id	FullDescription	Bedrooms	City	State	Prediction
40	39	A reasonable sized house having three bedrooms with four bathrooms and one visitors toilet in Scottsdale city, Maricopa County. AZ	3	Scottsdale	AZ	1142046
41	40	A reasonable sized house having three bedrooms with three bathrooms in Paso Robles city, San Luis Obispo County. CA	3	Paso Robles	CA	586765.5
42	41	A big house having four bedrooms with one bathroom in Bothell city, Snohomish County. WA	4	Bothell	WA	629833.8
43	42	A big house having five bedrooms with two bathrooms in Bothell city, Snohomish County. WA	5	Bothell	WA	730601.9
44	43	A reasonable sized house having three bedrooms with two bathrooms in Bothell city, Snohomish	3	Bothell	WA	463773.2
45	44	A reasonable sized house having three bedrooms with two bathrooms and one visitors toilet in Bothell city, Snohomish County. WA	3	Bothell	WA	445307.8
46	45	A small sized flat having four bedrooms with two bathrooms in Loma city, Mesa County. CO	4	Loma	CO	361928.7
47	46	A reasonable sized house having three bedrooms with three bathrooms in Loma city, Mesa County. CO	3	Loma	CO	530519.4
48	47	A reasonable sized house having four bedrooms with three bathrooms in Loma city, Mesa County. CO	4	Loma	CO	531854.2

Houses_Dataset.zip_Nystroem_Ker   ⊕

Table 11.15 – A prediction table from a multimodal dataset

The **Prediction** column presents the predicted price for each row. This finalizes the process of making predictions with multimodal datasets. As expected, after models made from multimodal datasets have been deployed, predictions can be made against them.

Summary

In this chapter, we have explored how insights can be generated from an image, location, and free text. In so doing, we highlighted the benefits that these data types present, as well as the challenges that come with each of them. We also pointed out how these are typically addressed in the mainstream. We proceeded to build models with a multimodal dataset using DataRobot and make predictions from the model. We also looked at a variety of ways to derive insights from the location, free text, and image aspects of the models. By demonstrating the process of model building using a multimodal dataset, we showed how DataRobot simplifies the handling of the challenges different data types pose.

Having said that, it is important to draw attention to the fact that DataRobot appears to have some limitations in terms of free text processing. Whilst the platform significantly simplifies the process of text processes, at the time of this publication, we are unsure of the extent to which domain-specific stop words can be included in the DataRobot process. It appears generic stop words are dropped, but sometimes there are domain-specific stop words that need to be accounted for. Elsewhere, within the context of multimodal modeling, we are unsure whether the text aspects of models could be tuned to include and alter the methods of stemming and lemming. It is therefore recommended that you perform your own text processing and feature engineering before feeding text into DataRobot to achieve better results.

In this chapter, as well as the previous ones, we have interfaced with DataRobot using the platform. Although the platform comes with numerous capabilities, these capabilities come with some limitations. These limitations, together with how they can be alleviated using programmatic access to the platform, are extensively covered in *Chapter 12, DataRobot Python API*.

12
DataRobot Python API

Users can access DataRobot's capabilities using DataRobot's Python client package. This lets us ingest data, create machine learning projects, make predictions from models, and manage models programmatically. It is easy to see the advantages that **Application Programming Interfaces (APIs)** offer users. The integrated use of Python and DataRobot lets us leverage the AutoML capabilities DataRobot presents, all while exploiting the programmatic flexibility and potential that Python possesses.

In this chapter, we will use the DataRobot Python API to ingest data, create a project with models, evaluate the models, and make predictions against them. At a high level, we will cover the following topics:

- Accessing the DataRobot API
- Understanding the DataRobot Python client
- Building models programmatically
- Making predictions programmatically

Technical requirements

For the analysis and modeling that will be carried out in this chapter, you will need access to the DataRobot software. Jupyter Notebook is crucial for this chapter as most of the interactions with DataRobot will be carried out from the console. Your Python version should be 2.7 or 3.4+. Now, let's look at the dataset that will be utilized in this chapter.

Check out the following video to see the Code in Action at `https://bit.ly/3wV4qx5`.

Automobile Dataset

The automobile dataset can be accessed at the UCI Machine Learning Repository (`https://archive.ics.uci.edu/ml/datasets/Automobile`). Each row in this dataset represents a specific automobile. The features (columns) describe its characteristics, risk rating, and associated normalized losses. Even though it is a small dataset, it has many features that are numerical as well as categorical. Its features are described on its web page and the data is provided in `.csv` format.

> **Dataset Citation**
>
> Dua, D. and Graff, C. (2019). UCI Machine Learning Repository (`http://archive.ics.uci.edu/ml`). Irvine, CA: University of California, School of Information and Computer Science.

Accessing the DataRobot API

The programmatic use of DataRobot enables data experts to leverage the platform's efficacies while having the flexibility associated with typical programming. With the API access of DataRobot, data from numerous sources can be integrated for analytic or modeling purposes. This capability is not only limited to the data that's ingested, but also the output of the outcome. For instance, API access makes it possible for a customer risk profiling model to get data from differing sources, such as Google BigQuery, local files, as well as AWS S3 buckets. And in a few lines of codes, the outcomes can update records on Salesforce, as well as those surfaced on PowerBI via a BigQuery table. The strength of this multiple data source integration capability is furthered as this enables the automated, scheduled, end-to-end periodic refresh of model outcomes.

In this preceding case, it becomes possible for the client base to be rescored periodically. Regarding scoring data, the DataRobot platform can only score datasets that are less than 1 GB in size. When problems require huge datasets, the **Batch Prediction API** normally chunks up the data and scores them concurrently. For a dataset with hundreds of millions of rows, it is possible to set up an iterative job to chunk up the data and score it iteratively using the Batch Prediction API.

In addition, the API access to DataRobot allows users to develop user-defined features that make commercial sense before analysis and those based on scored model outcomes. This makes the modeling process more robust as it allows human intelligence to be applied to outcomes. In the preceding client risk profiling case, it becomes possible to classify customers into risk categories for easier business decision making. Also, based on the explanations given, the next best actions could be developed.

Furthermore, programmatic use of DataRobot allows users to configure differing visualizations as they deem fit. This also offers analysts a broader range of visual outcome types. The Seaborn and Matplotlib Python libraries offer a huge range of visualization types with differing configurations. This also allows certain data subgroups or splits to be visualized. Among other benefits, it becomes possible to even select certain aspects of the data to be visualized.

One of the big advantages of accessing DataRobot using its API is the ability to create multiple projects iteratively. Two easy examples come to mind here. One approach to improving the outcomes of multi-class modeling is to use the one versus all modeling paradigm. This involves creating models for each of the classes. When scoring, all the models are used to score the data and for each row, the class with the highest score is attributed to the row. To bring this to life, let's assume we are building models to predict wheel drive types based on other characteristics. First, models are created for the three main types of wheel drives; that is, **front-wheel drive (FWD)**, **four-wheel drive (4WD)**, and **rear-wheel drive (RWD)**. Data is then scored against all three models, and the model that presents each row with the highest prediction is assumed as the class the row belongs to.

The model factory is another example where multiple model projects are integrated into a system so that each project builds models for a subgroup in the data. In some problems, data tends to be nested in that certain variables tend to govern the way models behave generally. A point in case is modeling the performance of students nested in class. These features, such as the class teacher for schools, tend to control the effect other exogenous variables have on the dependent variable.

In the case of cars, their brands typically drive their prices. For instance, irrespective of how similar a Skoda is to an Audi, the Audi will most likely be more expensive. As such, when developing models for such a case, it is ideal to create models for each of the car brands. In the context of programmatically accessing DataRobot, the idea would be to run an iteration of the project for each of the car brands.

In addition to creating and scoring DataRobot models programmatically, we will use Jupyter Notebook's **Integrated Development Environment (IDE)** to build projects for a case of one versus all and a model factory. However, before we can create projects with DataRobot using an API, certain identification processes must be covered. Let's have a look.

To programmatically access DataRobot, users need to create an API key. This key is then used to access the platform from a client. To create an API key, open the **Account** menu at the top right-hand corner of the home page (see *Figure 12.1*). From there, access the **Developer Tools** window (see *Figure 12.1*):

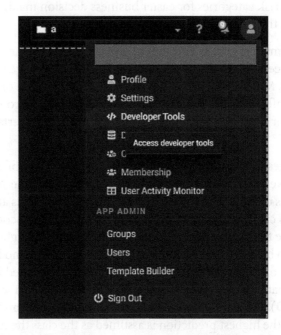

Figure 12.1 – Accessing Developer Tools

After opening the **Developer Tools** window, click on **Create New Key** and enter the name of the new key. On saving the new key's name, the API key will be generated (see *Figure 12.2*). After this, the generated key is copied and secured. The API key, along with the endpoint, is necessary to establish a connection between the local machine and the DataRobot instance:

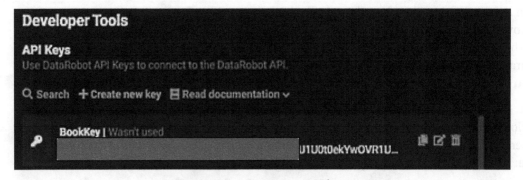

Figure 12.2 – Creating an API key

The endpoint parameter is the URL of the DataRobot endpoint. `https://app.datarobot.com/api/v2` is the default endpoint for the US cloud-based endpoint for its US and Japanese users. The EU-managed cloud endpoint is `https://app.eu.datarobot.com/api/v2`. VPC, on-premises, hybrid, or private users usually have their deployment endpoint as their DataRobot GUI root. To enhance security, these credentials are sometimes stored and accessed as `.yaml` files. These two credentials enable a connection between a computer and a DataRobot instance to use the DataRobot Python client.

Using the DataRobot Python client

The Python programming language is one of the most popular programming languages used by data scientists. It is flexible yet powerful. Being able to integrate the AutoML capabilities of DataRobot and utilize the flexibility of Python offers data scientists various benefits, as we mentioned earlier.

Programming in Python using the Jupyter IDE.

Now, let's explore the DataRobot Python client.

To use the DataRobot Python client, Python must be version 2.7 or 3.4+. The most up-to-date version of DataRobot must be installed. For the cloud version, the `pip` command will install the most recent version of the `DataRobot` package. On Python, running `!pip install datarobot` should install the `DataRobot` package.

Having installed the `DataRobot` package, the package has been imported. The `Client` method of the `DataRobot` package provides the much-needed connection to the DataRobot instance. As shown in *Figure 12.3*, the basic format for the `Client` method is as follows:

```
Import DataRobot as dr
dr.Client(endpoint= 'ENTER_THE_ENDPOINT_LINK', token = 'ENTER_
YOUR_API TOKEN')
```

In terms of data ingestion, data can be imported from different sources. This process is identical to normal data imports with Python. The local file installation is quite straightforward. Here, all you need is the API key and the file path. *Figure 12.3* presents the code for ingesting the automobile dataset. For the JDBC connection, to get data from platforms such as BigQuery and Snowflake, in addition to the API key, the identity of the data source object is required, as well as the user database's credentials – their usernames and passwords. The user database's credentials are provided by their organization's database administrators.

In this section, we established how to access the credentials necessary to programmatically use DataRobot. We have also imported data programmatically. Naturally, conducting some analysis and modeling comes after ingesting data. In the next section, we will create machine learning models using the Python API.

Building models programmatically

Now that we have imported the data, we will start building models programmatically. We will look at building the most basic models, then explore how to extract and visualize feature impact, before evaluating the performance of our models. Then, we will create more complex projects. Specifically, we will build one versus all **multiclass** classification models and **model factories**.

To create a DataRobot project, we must use the DataRobot `Project.start` method. The basic format for this is importing the necessary libraries (DataRobot, in the following case). Thereafter, the access credentials are presented, as described in the previous section. It is at the point that the `Project` method is called. `project_name`, `sourcedata`, and `target` are the minimal parameters that are required by the `Project` method for projects to be created. The `project_name` parameter tells DataRobot the name to give the created project. `sourcedata` provides information regarding the location of the data that's required to create models. This could be a location or a Python object. Finally, `target` specifies the target variable for the models to be built, as shown here:

```
import datarobot as dr
dr.Client(endpoint= 'ENTER_THE_ENDPOINT_LINK', token = 'ENTER_
YOUR_API TOKEN')
project = dr.Project.start(project_name = 'ENTER_PROJECT_NAME',
sourcedata='ENTER_DATA_LOCATION',
target='ENTER_YOUR_TARGET_VARIABLE')
```

The basic format for creating projects was shown in the preceding section and illustrated in *Figure 12.3*. Once the models have been created, we can use the `project.get_models` method to get a list of them. This list of models is presented in order by their validation scores by default. For this example, we will be using the automobile dataset, which we used to build models in *Chapter 6, Model Building with DataRobot*. The project's name is `autoproject_1`. Here, the file's location is specifically stored in a pandas object called `data`. The target variable is `price`. Note that these parameters are case-sensitive:

```
#Import os, pandas as pd and the DataRobot Package as dr then set the dr Client
#Create the project
import os
import datarobot as dr
import pandas as pd
import matplotlib.pyplot as plt
import seaborn as sns

dr.Client(endpoint= 'https://app.eu.datarobot.com/api/v2',
          token = 'Nj                                                    A9')
df = pd.read_csv('C:\                                       \Automobile.csv')

project_1 = dr.Project.start(project_name = 'autoproject_1',
                             sourcedata=df,
                               metric = 'RMSE',

                               target='price')
```

```
project_1.get_models()[0:10]
```

```
[Model('Gradient Boosted Greedy Trees Regressor (Least-Squares Loss)'),
 Model('ENET Blender'),
 Model('ENET Blender'),
 Model('AVG Blender'),
 Model('eXtreme Gradient Boosted Trees Regressor (Poisson Loss)'),
 Model('eXtreme Gradient Boosted Trees Regressor (Poisson Loss)'),
 Model('Gradient Boosted Trees Regressor (Least-Squares Loss)'),
 Model('Advanced AVG Blender'),
 Model('eXtreme Gradient Boosted Trees Regressor (Poisson Loss)'),
```

Figure 12.3 – Programmatically creating DataRobot models and extracting their lists

Once you've created the model, the get_models method is called to list the models. We can see that the best performing model is Gradient Boosted Greedy Trees Regressor (Least-Square Loss). To evaluate this model, we need to extract its ID. To do so, we must create an object, best_model_01, to store the best-performing model. This metrics method is then called for this model. As shown in the following screenshot, the cross-validation RMSE for this model is 2107.40:

```
In [128]:  #Pick best model
           best_model_1 = project_1.get_models()[0]

           print(best_model_1) #Print best model's name
           best_model_1.metrics['RMSE']['crossValidation'] #Print crossValidation score
```

```
           Model('Gradient Boosted Greedy Trees Regressor (Least-Squares Loss)')
```

```
Out[128]:  2107.401684
```

Figure 12.4 – Programmatically evaluating DataRobot models

To provide some insight into the price drivers, we need the feature impacts. These can be retrieved through the DataRobot API using the `get_or_feature_impact` method. To visualize the feature impacts for projects, we must define a function called `plot_FI` that takes in the model's name and chart title as parameters, gets the feature impacts, and then normalizes and plots them using Seaborn's bar plot method. Regarding the `autoproject_1` project, the following screenshot shows how to retrieve and present the feature impacts using the `plot_FI` function:

```
In [156]:  def plot_FI(model, title=None):
               #Get feature impact
               feature_impacts = model.get_or_request_feature_impact()

               #Sort feature impact based on normalised impact
               feature_impacts.sort(key=lambda x: x['impactNormalized'], reverse=True)

               fi_df = pd.DataFrame(feature_impacts) #Save feature impact in pandas dataframe
               fig, ax = plt.subplots(figsize=(14,5))
               b = sns.barplot(x="featureName", y="impactNormalized", data=fi_df[0:5],
                               color="b")
               b.axes.set_title('Feature Impact' if not title else title,fontsize=20)

           plot_FI(best_model_1, title = 'Drivers of Price')
```

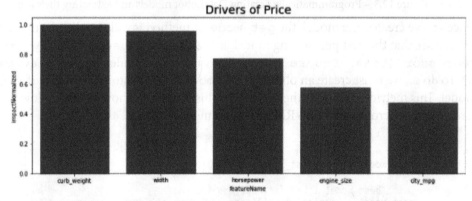

Figure 12.5 – Defining a function and extracting the feature impacts

Programmatic access to DataRobot furthers the benefits the platform offers. With programmatic access, you can take advantage of the iterative process within Python, and users can create multiple projects for the same dataset. Now, let's look at two ways to create multiple projects from the same dataset: **multi-class** classification and **model factory**.

Multi-class classification involves classifying instances into more than two classes. It is possible to create a single project that classifies rows into either of these classes. Essentially, this is a model that classes rows into one of all the available classes. Another way to approach this problem involves building different models for the different classes. Within this approach, a model is built for each of the classes as a target. You can see how this can be executed using Python's iterative process; that is, by looping through all the target levels. The one versus all method is better for performing classification problems with more than two classes.

Now, let's demonstrate how to use the one versus all method on the auto pricing project. Here, we will create price classes using the pandas **quantile-base discretization** function, qcut. qcut helps divide data into similarly sized bins. Using this function, we can divide our data into price classes – low to high. The following screenshot shows this price discretizing process and checking the distribution of cases across the classes:

```
In [94]:  #Split price into quartiles
          bin_labels = ['Low', 'Mid_1', 'Mid_2', 'High']
          df['price_class'] = pd.qcut(df['price'],
                                      q=[0, .25, .5, .75, 1],
                                      labels=bin_labels)
          df.head()
          df['price_class'].value_counts()
          #one vs all

Out[94]:  Low      53
          High     51
          Mid_2    51
          Mid_1    50
          Name: price_class, dtype: int64
```

Figure 12.6 – Price discretization

Having created the classes, to allow for data **leakages**, we will drop the initial price variable. We will write a loop that builds models for each of the price classes. Perform the following steps:

1. Turn the price_class variable into dummy variables.

2. For each iteration, create a DataRobot project after a dummified price class name.

3. For each iteration, we drop the price_class dummy level being modeled. This ensures that there are no leakages.

4. For each iteration, we must build the models for a target variable dummy.

5. After creating the projects, the top-performing model for each project is selected and stored in a dictionary:

```python
targetVars = pd.get_dummies(df['price_class'], prefix = 'price_class').columns.to_list()
df = pd.get_dummies(df.drop(['price'], axis=1), columns=['price_class'])

def dr_train(data,target):
    proj = dr.Project.create(data,project_name='Auto_'+target,max_wait=3600)
    proj.set_target(target=target,mode=dr.enums.AUTOPILOT_MODE.FULL_AUTO,max_wait=3600)
    return proj.id
```

```python
projInfo = pd.DataFrame()
for i in targetVars:
    data = df.drop([elem for elem in targetVars if elem not in i],axis=1)
    proj_id = dr_train(data=data,target=i)
    projInfo = projInfo.append({"target":i,"proj_id":proj_id},ignore_index=True)
```

```python
# This is for referencing the projects if we created them previously

# Get projects if already created above.
all_my_projects = dr.Project.list()
Auto_projects = [p for p in all_my_projects if 'Auto_' in p.project_name]

# Collect each project in a dataframe
projects = {}

for p in Auto_projects:
    auto_class = p.project_name.split('Auto_')[1]
    projects[auto_class] = p

# Collect the preferred model from each project in a dictionary
most_accurate_models = {}

# Retrieve the top models (by cross-validation score)
for p, project in projects.items():
    most_accurate = project.get_models()[0]
    most_accurate_models[auto_class] = most_accurate
```

Figure 12.7 – Creating a one versus all classification suite of projects

This process involves creating projects with a suite of models with targets iterating through all the price classes. After creating the projects, the best model for each target class is selected using an iteration of all the projects with names starting with Auto, and then the top-performing model for each project. These best models are placed in a dictionary.

It is sometimes recommended, if not ideal, to create different projects with a subset of the data. After selecting all the cases for the target variable, you must create a random subset of the data for each project creation iteration. In the auto pricing case, however, we were unable to explore this as the out-sample size was limiting.

A **model factory** is a multi-level modeling system where a model is developed for a subgroup of cases. For instance, the price of a car might be heavily determined by its fuel type in that it becomes beneficial to build different models for each fuel type within the same system. Programmatically building model factories is somewhat like the one versus all approach to classification. Instead of projects being iteratively created for each of the unique target variable levels, as with the one versus all process, the model factory involves building models for each level of a predictor variable. The key steps in building the model factory process, which involves iterating through each of the unique variable levels, are as follows (`fuel-type`):

1. First, create and store a project.

2. Select cases for the target variable (the influencer of interest). In this case, the variable is `fuel-type`. Here, this variable is selected, and differing levels of this variable are used to create DataRobot projects. In simple terms, this step involves, for instance, selecting all the rows with `fuel-type` set to `gas` as a subgroup.

3. If necessary, define the evaluation metric. Here, we can alter aspects of the advanced options we encountered in *Chapter 6, Model Building with DataRobot*. Other advance options can be selected and altered.

4. If necessary, set a data limit that a class will be deselected for (for instance, if the number of rows is less than 20 for that class). The importance of this step lies in the fact that some variable levels could have very low occurrences, so the sample size within the subgroup is small. Therefore, creating models out of these becomes a challenge. This step becomes the best place to drop such variable levels using the count of cases within the subgroup.

5. All the models from all the projects are selected and stored in a dictionary.

Some of these steps are evident in creating a model factor for the auto pricing problem (see *Figure 12.8*). Here, fuel-type is selected as the feature that projects are created on. In this case, only two projects are created: one for gas automobiles and another for diesel ones. Now that we've created the models, the next step is to collect the best-performing models for each fuel-type:

```
projects = {} #To save projects

#Create one project for each fuel-type type
for value in df['fuel-type'].unique():

        temp_project = dr.Project.start(df.loc[df['fuel-type'] == value], #create a project
                                project_name = 'price_'+value, #name the project
                                target = 'price', #set the target variable
                                metric = 'RMSE')
        projects[value] = temp_project #collect the project
        pass
#Wait for all autopilots to finish
for key in projects:
    log = projects[key].wait_for_autopilot()
```

Figure 12.8 – Creating model factories

The efficacy of using one versus all multiclass classification models and model factories lies in their ability to fit models to each level of the target variable. This happens in an automated fashion and considers the sample validation, all the preprocessing steps, and the model training process. When data cardinality and volume are high, these approaches would mostly outperform typical modeling.

For the model factory, multiple projects are created for the different levels of the feature of interest. To evaluate this, the best-performing model for each project is selected from the dictionary for all projects. This set of best models from all the projects is stored in another dictionary object. A `for` loop is then run across all the models of the dictionary to extract the performance of the model, as shown in the following screenshot:

```
best_models = {} #To save models
for key in projects:
    best_models[key] = projects[key].get_models()[0]
    print('-----------------------------------')
    print('Best model for admission type id: %s' %key)
    print(best_models[key])
    print(best_models[key].metrics['RMSE']['crossValidation'])
    print('-----------------------------------')
```

```
-----------------------------------
Best model for admission type id: gas
Model('ENET Blender')
2073.556834
-----------------------------------
-----------------------------------
Best model for admission type id: diesel
Model('Ridge Regressor')
2415.255848
-----------------------------------
```

Figure 12.9 – Evaluating the performance of models with a model factory

Improved model performance is only one of the reasons you should use the one versus all multiclass classification models, as well as model factors. Sometimes, understanding the drivers is equally as important. Visualizing the feature importance for the different fuel types could present an interesting contrast in drivers. This means that different factors affect the prices of different fuel types. This could have a bearing on strategic decisions. As shown in the following screenshot, the Python API can be used to plot the feature impacts by leveraging chart functions from Seaborn and Matplotlib:

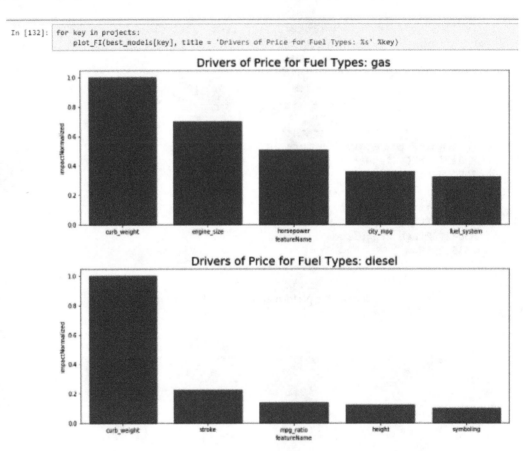

Figure 12.10 – Feature impacts for the differing diesel and gas automobiles

As we can see, there are some differences in the feature impacts for the automobile fuel types. While `curb-weight` seems to be an important driver, its effect is relatively more important for diesel vehicles. Similarly, for gas cars, the power that's generated by these automobiles, as typified by the `engine_size` and `horsepower` features, carries more importance in determining price than those of diesel cars. You can already see the effect such preliminary findings could have on decisions and how this could be applied to other commercial cases. Using feature importance to examine multiple models can also be applied in the case of one versus all classification problems.

In this section, we created basic DataRobot projects using the Python API. After, we solved more complex problems by using multiple projects within a system. There, we created one versus all projects to solve multiclass classification problems and model factories to solve multi-level problems involving subgroups. We also explored feature impact and model evaluation. Having programmatically created models, we will now learn how to make predictions using these models. Specifically, we will learn how to deploy models, make predictions, extract explanations from models, and score large datasets through parallelization.

Making predictions programmatically

The possibilities that programmatically using DataRobot presents are enormous. By using its API, models can be deployed and predictions can be made against them. Before making programmatical predictions within the production environment, models need to be deployed. DataRobot models are deployed using Portable Prediction Servers. These are Docker containers that can host machine learning models, which serve predictions and prediction explanations through a REST API.

To deploy models, we can use the DataRobot package's `deployment` method. Here, we must provide a description, the DataRobot model's ID, as well as its label to create the deployments. A typical Python deployment script follows this format:

```
deployment = dr.Deployment.create_from_learning_model(
MODEL_ID, label='DEPLOYMENT_LABEL', description='DEPLOYMENT_
DESCRIPTION',
    default_prediction_server_id=PREDICTION_SERVER_ID)
deployment
```

As per this approach, the following screenshot shows how `autoproject_1`, which we created in the *Building models programmatically* section, can be deployed. Here, the model ID is `best_model_1`. We will label `AutoBase Deployment` with a description of `Base Automobile Price Deployment`:

```
In [152]:   PredictionServ = dr.PredictionServer.list()[0]

            deployment = dr.Deployment.create_from_learning_model(
                best_model_1.id, label='AutoBase Deployment',
                description='Base Automobile deployment',
                default_prediction_server_id=PredictionServ.id)
            deployment

Out[152]:   Deployment(AutoBase Deployment)
```

Figure 12.11 – Deploying a model programmatically

The deployment process can be iterated to enable those of more complex projects. For instance, with model factories, irrespective of the number of levels the differentiating variable has, with a single `for` loop, all the best models can be deployed to DataRobot. For each of the best models, a deployment is created, which is then used to score new data. The script for deploying the model factory for the automobile project, along with the fuel type as its differentiating variable, is shown in the following screenshot:

```
PredictionServ = dr.PredictionServer.list()[0]

for key in best_models:
    autoMF_deployment = dr.Deployment.create_from_learning_model(
                        best_models[key].id, label='Automobile Prices: %s' %key,
                        description='Auto deployment',
                        default_prediction_server_id=PredictionServ.id
                        )
```

Figure 12.12 – Deploying models from a model factory

Having deployed the models, predictions can be made against them. To make simple predictions within the development environment, we can use the `DataRobot BatchPredictionJob.score_to_file` method. To make predictions, this method requires the model ID, prediction data, and the location where the scored data will be stored. Here, we will use `best_model_1` to score the same model we used to develop the model, the `df` data object, and the location path, which specifies the prediction file path as `./pred.csv`. The `passthrough_columns_set` parameter specifies the columns from the original dataset that will be included in the predictions. Since this is set to `'all'`, all the columns are returned, as shown here:

```
prediction = dr.BatchPredictionJob.score_to_file(
    deployment.id,
    df,
    './predicted.csv',
    passthrough_columns_set='all')

prediction
```

```
BatchPredictionJob(batchPredictions, '6▮▮▮▮▮▮▮▮▮▮▮▮▮▮▮▮', status=INITIALIZING)
```

Figure 12.13 – Simple programmatic prediction

These predictions comprise all the columns from the initial dataset, in addition to the predicted prices. There are cases where it is ideal to include rationales behind predictions. In such cases, the max_explanations parameter should be included in the job's configuration. This parameter sets the highest number of explanations to be provided for every data row.

Summary

DataRobot provides us with a unique capability to rapidly develop models. With this platform, data scientists can combine the benefits of DataRobot and the flexibilities of open programming. In this chapter, we explored ways to access the credentials needed to programmatically use DataRobot. Using the Python client, we demonstrated ways in which data can be ingested and how basic projects can be created. We started building models for more complex problems. We created model factories as well as one versus all models. Finally, we demonstrated how models can be deployed and used to score data.

One of the key advantages of programmatically using DataRobot is the ability to ingest data from numerous sources, score them, and store them in the relevant sources. This makes it possible to carry out end-to-end dataset scoring. It becomes possible for a system to be set up to score models periodically. With this comes numerous data quality and model monitoring concerns. The next chapter will focus on how to control the quality of the models and data on the DataRobot platform, as well as using the Python API.

13
Model Governance and MLOps

In the previous chapters, we learned how to build, understand, and deploy models. We will now learn how to govern these models and how to responsibly use these models in operations. In earlier chapters, we discussed the methods for understanding the business problem, the system in which the model will operate, and the potential consequences of using a model's predictions. **MLOps** is a word made up of **machine learning and DevOps**. It is made of processes and practices to efficiently, reliably, and effectively operationalize the production of **machine learning** (**ML**) models within an enterprise. MLOps aims to ensure commercial value and regulatory requirements are met continuously by ensuring production models' outcomes are of good quality and automation is in place. It provides a centralized system to manage the entire life cycle of all ML models in production.

Activities within MLOps cover all aspects of model deployment, provide real-time tracking accuracy of models in production, offer a champion challenger process that continuously learns and evaluates models using real-time data, track model bias and fairness, and provide a **model governance** framework to ensure that models continue to deliver business impact while meeting the regulatory requirements. In *Chapter 8*, *Model Scoring and Deployment*, we covered model deployment on the DataRobot platform.

Furthermore, in *Chapter 8*, *Model Scoring and Deployment*, we extensively covered aspects of monitoring models in production. Given the crucial role model governance plays within the MLOps process, in this chapter, we will introduce the model governance framework. One key aspect of model monitoring is to ensure that models are not biased and are fair towards all people impacted by the model, which we will explore in this chapter. After that, we will take a deeper look at how to enable other aspects of MLOps, including how to maintain and monitor models. Thus, we're going to cover the following main topics:

- Governing models
- Addressing model bias and fairness
- Implementing MLOps
- Notifications and changing models in production

Technical requirements

Most parts of this chapter require access to the DataRobot software. The example utilizes a relatively small dataset, **Book-Crossing**, consisting of three tables, whose manipulation was described earlier in *Chapter 10*, *Recommender Systems*. As will be covered in the data description, we will create new fields in addition to those used in *Chapter 10*, *Recommender Systems*.

Book-Crossing dataset

The example used to illustrate the aspects of model governance is the same as the one used for building recommendation systems in *Chapter 10*, *Recommender Systems*. The dataset is based on the Book-Crossing dataset by Cai-Nicolas Ziegler and colleagues (`http://www2.informatik.uni-freiburg.de/~cziegler/BX/`). The data was collected during a 4-week crawl from the Book-Crossing community between August and September 2004.

> **Important Note**
>
> Before using this dataset, the authors of this book have informed the owner of the dataset about its usage in this book:
>
> *Cai-Nicolas Ziegler, Sean M. McNee, Joseph A. Konstan, Georg Lausen (2005). Improving Recommendation Lists Through Topic Diversification. Proceedings of the 14th International World Wide Web Conference (WWW 2005). May 2010–2014, 2005, Chiba, Japan*

The subsequent three tables, provided in `.csv` format, make up this dataset.

- Users: This table presents the profile of the users, with anonymized `User-ID` and presented as integers. Also provided are the user `Location` and `Age`.

- Books: This table contains the characteristics of the books. Its features include `ISBM`, `Book-Title`, `Book-Author`, `Year-Of-Publication`, and `Publisher`.

- Ratings: This table shows the ratings. Each row provides a user's rating for a book. The `Book-Rating` is either implicit as `0` or explicit between `1` and `10` (the higher, the more appreciated). However, within the context of this project, we will focus solely on ratings that are explicit for the model development. The table also includes the `User-ID` and `ISBN` fields.

In addition, using Excel, we created two extra fields using age and a rating column. We created the `RatingClass` field, which considers a rating over 7 as a `High` rating or else it is `Low`. Similarly, we created the `AgeGroup` field; this classes ages over 40 as `Over Forty` and those under 25 as `Under 25`, or else they are considered simply `Between 25 and 40`. Finally, we dropped out data rows with a missing age column.

Governing models

Organizations using ML governance define a framework of rules and controls for managing the ML workflows pertaining to model development, production, and post-production monitoring. The commercial importance of ML is well established. Still, only a fraction of companies investing in ML are realizing the benefits. Some establishments have struggled to ensure that the outcomes of ML projects are well aligned with their strategic direction. Importantly, many organizations are subject to regulations, such as the recently implemented General Data Protection Regulation within the European Union and European Economic Area, which affect the use of these models and their outputs. Businesses, in general, need to steer their ML use to ensure regulatory requirements are satisfied and strategic goals and values are continually realized.

Having an established governance framework in place ensures that data scientists can focus on the innovative part of their role, which is solving new problems. With governance, data scientists spend less time assessing the commercial value their models are delivering to the business, evaluating models' performance of production, and examining whether there has been data drift. Model governance simplifies the model versioning and change tracking process for all production models. This is always a key aspect of ML audit trailing. In addition, notifications could be set up to alert stakeholders when a model in production encounters anomalies and changes in performance. When there is a significant decline in performance, models in production could be swapped with better-performing challenger models in a seamless fashion. Although this process might require reviews and authorization from other stakeholders, it is much more simple and straightforward than a typical data science workflow.

It is clear that governing models throughout the entire process is a complex and time-consuming undertaking. Without tool support, it is easy for the data science teams to miss key steps. Tools such as DataRobot make this task easier, ensuring that many required tasks are performed automatically. This ease of use can also sometimes make the teams use these tools without thinking. This can be dangerous too. Thus, a judicious mix of using the tools such as DataRobot and setting up process controls and reviews is needed to ensure proper governance.

DataRobot's MLOps provides organizations with an ML model governance framework that helps in the management of risks. Using the model governance tool, a business executive can track important business metrics and ensure that regulatory requirements are met on a continuous basis. They can easily assess the model performance in production to ensure that models are fit for purpose. Furthermore, with governance in place, the commercial criticality of models is defined before deployment. This ensures that when models are critical to the business, certain changes to the model need to be reviewed and authorized by stakeholders before such changes are fully implemented. In line with ethics, the use of ML models is expected to enable a fair process. So, models' outputs should be purged of any form of biases. In subsequent sections in this chapter, in addition to other aspects of MLOps, we will examine how bias could be mitigated in the ML model in development as well as in production.

Addressing model bias and fairness

A key characteristic of ML lies in its learning from the past to predict the future. This implies that future predictions would be influenced by the past. Some training datasets are structured in ways that could introduce bias into ML models. These biases are based on unspoken unfairness evident in human systems. Bias is known to maintain prejudice and unfairness that preexisted the models and could lead to unintended consequences. An AI system that is unable to understand human bias mirrors, if not exacerbates, the bias present in the training dataset. It is easy to see why women are more likely to receive lower salary predictions by ML models than their male counterparts. In a similar example, credit card companies using historic data-driven ML models could be steered into offering higher rates to individuals from minority backgrounds. Such **unwarranted associated** are caused by human bias that is inherent in the training dataset. It is unfair to include bias-laden features with unbiased ones in model development. A fair process considers an individual's payment history in making predictions about their credit, but unfair outcomes are possible when predictions are made based on the payment history of their family.

Supervised learning models can be particularly unfair, as certain data has circular dependency. For instance, to obtain a credit card, people need a credit history, and to have credit histories, credit cards are required. Since models are critical to credit assessment, it becomes nearly impossible for some people to get a credit card. Also, limited data about certain subgroups makes them more vulnerable to bias. This is because a minimal outcome distribution change in training data for such groups could skew the prediction outcomes for members of the group. These all point toward the extent to which ML models should manage bias and support a fair process.

Many industries – for instance, health care, insurance, and banking – take specific measures to guard against any form of bias and unfairness as a regulatory expectation. While it is inherently challenging to address bias in humans, it is somewhat easier to address ML bias. So, as part of ML governance, addressing ML bias could be pivotal in ensuring that their products don't amplify the skepticism about the ethical aspects of ML systems.

In addressing potential unwarranted outcomes, DataRobot has introduced a bias and fairness monitoring and control capability. This capability is selected and configured during model development. Let's step back and demonstrate how bias could be addressed in DataRobot. As with the typical platform, we upload the data as described in the preceding chapters. In the project configuration window (as within *Figure 13.1*), we open **Advanced Options** and the **Bias and Fairness** tab:

1. It is within this tab that we define protected features, how fairness is established and measured, as well as the target variable. We specify the fields in the prediction dataset that need to be protected. These are entered in the **Protected Features** input field. In this case, the `AgeGroup` field is selected as to be protected (see *Figure 13.1*). In some industry datasets, attributes such as sex, ethnicity, age, and religion must be selected. In this way, DataRobot manages and presents metrics to measure any potential model bias within each of the protected fields:

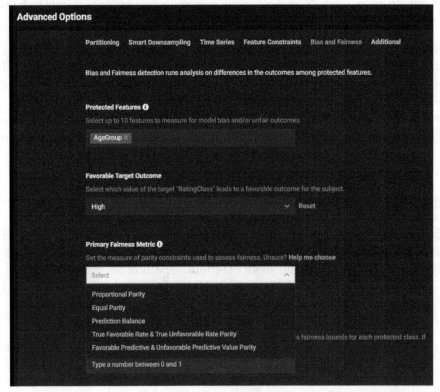

Figure 13.1 – Configuring bias and fairness during the model development

2. Next, the **Favorable Target Outcome** field is selected. This is the level of the target variable that is desirable. In this case, the target variable is the `RatingClass` level of `High`. This enables the measurement of bias on this level of the target variable.

3. The **Primary Fairness Metric** field outlines the metric against which fairness is measured. It is important to highlight that fairness differs considerably across use cases. The fairness for an insurance risk modeling tool would strive to ensure that the risk all potential clients carry is representative, whereas fairness within that employment tool would aim for members of a protected group to have similar chances of being employed when compared to those from other groups. The choice of **Primary Fairness Metric** helps DataRobot understand how to measure bias. A few metrics are presented to be selected. These include `Proportional Parity`, `Equal Parity`, `Prediction Balance`, `True Favorable Rate & True Unfavorable Rate Parity`, and `Favorable Predictive & Unfavorable Predictive Value Parity`.

4. If a user is unsure of the metric to choose, they can click on **Help Me Choose**, which presents a further set of questions. Answering these questions presents a recommendation of a **Fairness Metric** value, as shown in *Figure 13.2*:

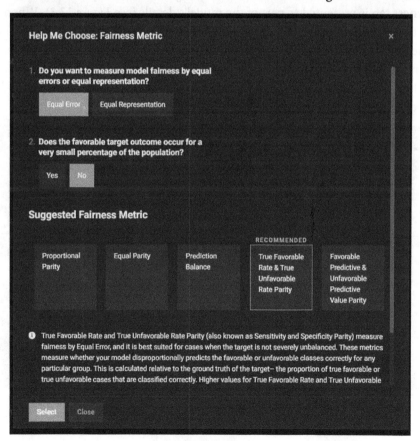

Figure 13.2 – Fairness Metric recommendation

In choosing our metric, because we are keenly interested in our model having similar prediction accuracy across age group membership, the **Equal Error** option is selected in response to how we want to measure model fairness. Since our outcome distribution is somewhat balanced between high and low, we choose **No** to the **Does the favorable target outcome occur for a very small percentage of the population?** question. Following this, DataRobot suggests **True Favorable Rate & True Unfavorable Rate Parity**. All throughout the process, the platform offers a description of the options and presents an explanation of the recommended metric, as well as those for other metrics.

5. A click on **Select** finalizes the process and the modeling process proceeds, as suggested in earlier chapters.

Model bias could be further examined after models have been developed. Since model bias and fairness were configured during model development, the **Bias and Fairness** tab is presented as part of the model's details (see *Figure 13.3*). When this tab is selected for any of the created models, the **Per-Class Bias** window is presented. Within this window, the relative extent to which the model is biased given the **Fairness Score** value is displayed. The `AgeGroup` **Per-Class Bias** value for the `Light Gradient Boosting on Elastic Predictions` model presented in *Figure 13.3* is below the default threshold:

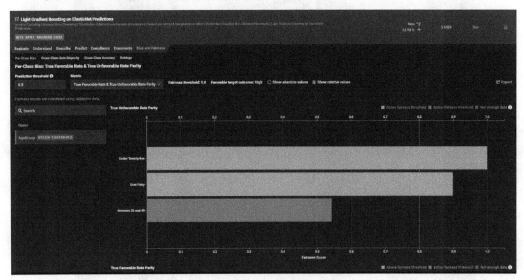

Figure 13.3 – Per-Class Bias exploration

According to this outcome, the accuracy of the model in predicting the true unfavorable outcome (of a Low rating) for individuals within the Between 25 and 40 class is lower than the other two classes. The score for this class falls below the default 80% threshold. The default threshold of 80% was applied for **Primary Fairness Metric**, as we didn't set a value for it during the model development, as seen in *Figure 13.3*. By clicking the **Show absolute values** tab, the absolute measures are presented instead. While the other chart (not visible in *Figure 13.3*) suggests that the accuracy for the favorable outcome was consistent across classes, this model could still be unfair, as it will most likely falsely predict unfavorable outcomes for individuals in the Between 25 and 40 class. *Figure 13.4* demonstrates how **Cross-Class Accuracy**, a set of more holistic accuracy metrics, could be used to assess accuracy across the protected classes:

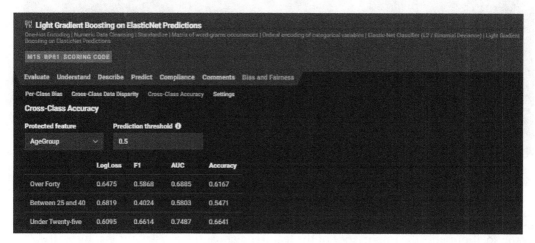

Figure 13.4 – Cross-Class Accuracy examination

Cross-Class Accuracy presents a set of accuracy metrics, assessing the model across differing levels of the `AgeGroup` class. As the outcome in *Figure 13.4* suggests, the accuracy of the model seems to be lower for the `Between 25 and 40` class across all accuracy measures. Because, as earlier alluded to, the performance of the model is similar across classes when it is the favorable class, only the lower true rate for the unfavorable outcome for the `Between 25 and 40` class seems to affect the fairness of the model. Because models learn from past data, exploring the features that might be responsible for this bias might be crucial in taking further actions. *Figure 13.5* shows the **Cross-Class Data Disparity** capability, which presents deep dives into why bias exists in ML models:

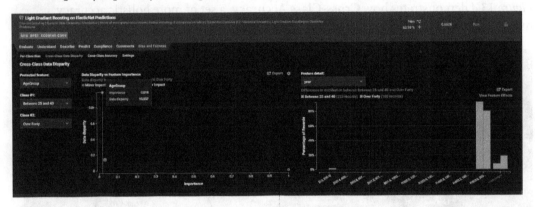

Figure 13.5 – A Cross-Class Data Disparity comparison between two age classes

To explore the rationale behind the model bias, the **Cross-Class Data Disparity** comparison compares the data distribution across two groups of a protected feature. In doing so, it presents the importance of the features against their distributional disparity. Of lowest importance, yet for obvious reasons with the largest disparity, the `Age-Group` feature seems to affect the model's accuracy. This is because `Age-Group`, being the predicted variable, will have the largest disparity in comparison to other variables, as it is identical to the predicted variable. The `year` book had a lower data disparity but had greater importance than the `Age-Group` feature. Further examination of the distribution of the year in the right-hand chart (*Figure 13.6*) shows that older books and books with a missing year seem to have been rated more by the `Over Forty` group in comparison with the `Between 25 and 40` group. On the contrary, the `Between 25 and 40` cohorts seem to be rated more of the newer books than their older counterparts.

When model bias exceeds an enterprise-established threshold, steps need to be taken to manage this unfairness. Options to address this unfairness include dropping features that might be responsible for the bias and retraining the model, or changing the model for a more ethical model. Most of the time, these changes ultimately affect the overall accuracy of the model. However, in our example case, `Light Gradient Boosting on Elastic Predictions` wasn't our best-performing model. DataRobot has within its bias and fairness toolkit the **Bias vs Accuracy** leaderboard comparison capability (see *Figure 13.6*):

Figure 13.6 – Bias vs Accuracy leaderboard

The **Bias vs Accuracy** chart assesses multiple models on their bias and accuracy. Here, we see that `Keras Residual AutoInt Classifier using Training Schedule` was the most accurate model and met the ethic threshold. In this case, this model could be deployed into production. It is important to note that neural network-based models are generally not accepted by many regulators today, but this could change in the future.

Processes concerning the assessment of ML model bias and fairness are expected to be integrated into the data science workflow to ensure model outcomes support a fair process. This becomes more important as conversations pertaining to ethical AI are becoming more ubiquitous across industries. Having looked at ways to ensure models are fair, we now progress into deploying the fair model, monitoring model performance in production, and other aspects of implementing MLOps in the next section.

Implementing MLOps

DataRobot, through its MLOps suite, provides capabilities to enable users to not only deploy models in production, but govern, monitor, and manage the models in production. In previous chapters, we have looked at how models are deployed on the platform and using the Python API client. MLOps provides an automated model monitoring capability, which tracks the service health, accuracy, and data drift of models in production. The automated real-time monitoring of production models ensures that models have high-quality outputs. Also, when there is a performance degeneration, stakeholders are notified, so action can be taken.

In this section, we will focus on aspects of model monitoring that we didn't cover in *Chapter 8*, *Model Scoring and Deployment*, of this book. We looked at how to examine the quality of deployment services, as well as changes in the underline feature distribution between the training and prediction data across time through the service health and data drift capabilities. As time passes, more recent data with target variables is introduced to the deployment. DataRobot can then examine models' initial predictions and establish models' actual accuracy in production. DataRobot also provides the capability to switch between alternative models in production. This section focuses on the evaluation of production model accuracy, setting up notifications, as well as switching models in production.

As you can guess by now, the job of the data science team does not end once a model is deployed. We now must monitor our models in production. After models have been deployed, before engaging in conversations pertaining to the monitoring of models, we need to control what individuals can do with those deployments. Stakeholder roles and responsibilities are important aspects of MLOps governance. Successful implementation of ML solutions depends on a clear definition of roles and what the actual duties of stakeholders are throughout ML models' production life cycles. As *Figure 13.7* highlights, when deployments are shared with other stakeholders, each stakeholder is given a role that defines their access level to that deployment:

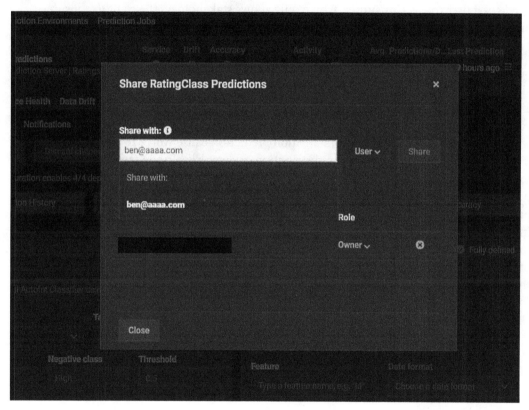

Figure 13.7 – Sharing deployments

To open the deployment sharing window (as shown in *Figure 13.7*), after the model was deployed, the deployment action button (the triple dash icon) on the top right-hand side was selected. Then, **Share** was chosen. Here, this **RatingClass Predictions** deployment was shared with a stakeholder, ben@aaaa.com. Importantly, this individual was given the role of **User**. With the **User** role, this stakeholder can write and read. In an actual sense, they can view the deployment, consume predictions, view deployment inventory, use the API to get data, and add other users to the deployment. The **Owner** level has additional administration rights and can perform business-critical operations, such as deleting the deployment, replacing the model, and editing the deployment metadata. The lowest user role is **Consumer**, which only allows stakeholders the right to consume predictions via the API route.

Production model monitoring ensures that models continue to deliver high-quality business impact as expected during development. A decline in this quality is a result of an alteration in the production data distribution or changes in the extent to which features affect the endogenous variable. For instance, changes in usage affect customer attrition, a variable of importance to a business. During a holiday period, the predictions for attrition would be higher. Such fluctuations in attrition prediction cause worry to the business if they are not expecting this change in distribution or data drift. In the same way, the extent to which predictor variables could influence a business outcome could also change. A point in case could be the effect of price on the propensity to buy. During the peak of a pandemic, individuals are far more conservative in their purchase of non-essentials. Now, imagine the chances of the accuracy of an in-production buying propensity model built for a non-essential product built before the pandemic. It is easy to see that the accuracy of the model will decay in production quite rapidly, thus having a significant impact on the business performance. Such situations raise the need to monitor the performance of models post-deployment.

In *Chapter 8, Model Scoring and Deployment*, we covered data drift, which examines changes in distribution between the training and production datasets, while accounting for their feature importance. Here, our focus will shift to monitoring the effect of variables on outcomes while in production. Changes in this effect could be established through the monitoring of production models' accuracy, a capability DataRobot offers. As part of the **Deployments Settings** window, as shown in *Figure 13.8*, there is an **Accuracy** tab:

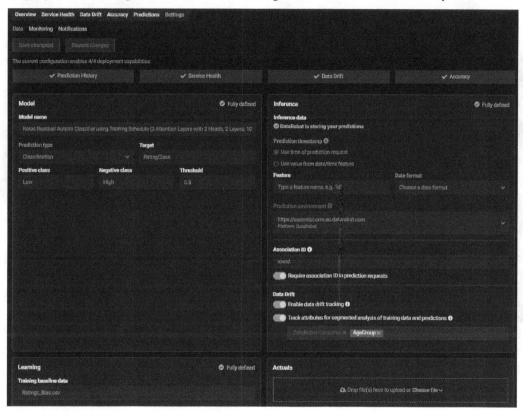

Figure 13.8 – Deployment window for accuracy setup

The **Accuracy** tab offers insight into the accuracy of production models. This capability allows users the ability to examine the performance of their production models over time. To compute the accuracy of a production model, actual outcomes need to be provided. After actuals have been uploaded, to generate accuracies, a set of fields needs to be completed. These include the **Actual Response** and **Association ID** fields, as well as those that are optional, **Was acted on** and **Timestamp** (see *Figure 13.9*):

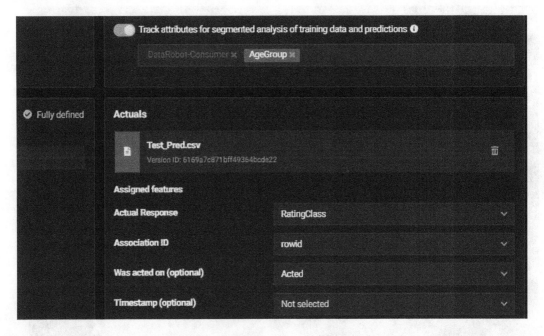

Figure 13.9 – Accuracy setup features

The **Actual Response** field specifies where the true outcome is in the data. In this case, the field is RatingClass. To link this to the earlier prediction dataset, **Association ID**, presented as rowid in this example, is requested to enable this connection. It is important to note that sometimes as a result of the models' predictions, action is taken by the business that could ultimately influence the outcome. To account for this possibility in calculating accuracy, the **Was acted on** and **Timestamp** variables are optionally requested (see *Figure 13.10* for the selection of these features):

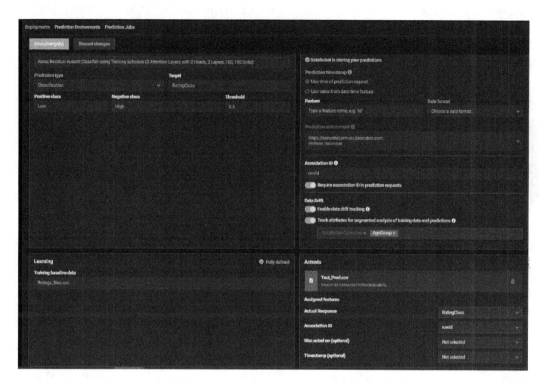

Figure 13.10 – Production accuracy identification feature selection

After the mandatory variables are selected, the **Save** button is clicked. This sets the computation off, thereafter opening the **Accuracy** window, displaying the production accuracy of the model. The performance of the production model is presented as tiles and as a graphical time series. *Figure 13.11* presents the **Deployment Accuracy** window. The **LogLoss, AUC, Accuracy, Kolmogorov-Smirmov,** and **Gini Norm** metrics tiles are selected. **Start** shows the model's performance against the holdout dataset during the development process. It appears that this model is better in production than during the training. Through the customize tiles, other metrics and their order could be chosen. The **Accuracy over Time** graph shows how the accuracy of the model has changed over time. The leftmost green spot on the graph indicates the model accuracy against the holdout dataset during development:

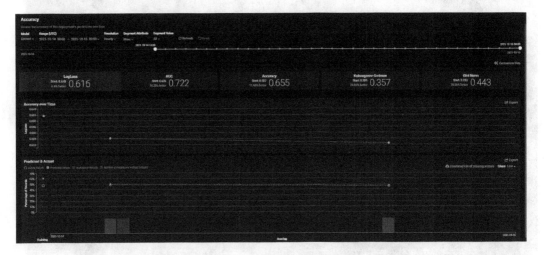

Figure 13.11 – production model performance assessment over time

The **Predicted & Actual** chart tells a similar story. Here, the selected class is Low. There is an option to change the class being explored. It is important to note that with these, the accuracy of the model on the differing levels of the AgeClass protected variable could be monitored. This could be done by selecting AgeClass in the **Segment Attribute** option and then choosing either of the levels in the **Segment Value** field. While in the present scenario production accuracy mirrors those of data drift, it is possible to configure notifications so that stakeholders are notified when metrics depart in a manner that adversely affects the business. In the next section, we will cover these notifications, as well as how to change models in deployment.

Notifications and changing models in production

In this chapter, we have established why the commercial impact of models can decay and ways to track this impact in the DataRobot platform. In cases where the end-to-end prediction process is fully automated and human intervention is limited, it becomes crucial that systems that notify stakeholders of any significant changes in the performance of production models are available. DataRobot can send notifications for significant changes in service health, data drift, and accuracy. These notifications can be set up and configured within the **Deployment** window:

1. From the **Settings** tab, select **Notifications**. As shown in *Figure 13.12*, three options are presented: notifications being sent for all events, notifications for critical events, and no notifications being sent. Notifications for all events are sent by email; all changes to the deployments are emailed to the owner:

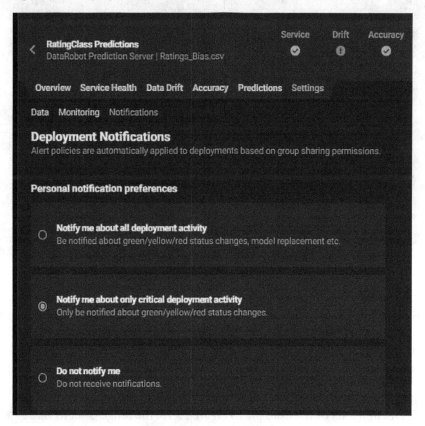

Figure 13.12 – Deployment Notifications setup

In *Figure 13.12*, notifications are set to notify me about only critical deployment activities. This setting implies that the stakeholders are notified when there are critical activities occurring on the deployment.

2. The **Monitoring** tab (in *Figure 13.13*) presents options for defining the notifications that are to be sent. Here, **Service Health** notification is set to be sent daily at 1:00. There are options to set notifications to occur anywhere between an hourly and quarterly monitoring cadence. When the box is unchecked, the notification is disabled:

Figure 13.13 – Monitoring notification setup

3. Notification for **Data Drift** has a few thresholds and configurations to be completed. As discussed in *Chapter 8*, *Model Scoring and Deployment*, data drift compares the distribution of incoming data to that used for the model development. Essentially, it looks at how recent production data differs from the training data across all features. Setting up a **Data Drift** notification involves the following considerations:

*   **Range** defines the period from which data is drawn to be compared with the development data. For the example in *Figure 13.13*, the range is set to Last 7 days, meaning that the data distribution for the preceding seven days is compared with that of the training data.

- Being a feature drift metric, the **Population Stability Index (PSI)** threshold defines the extent to which feature drift needs to occur for a notification to be triggered. Here, the threshold is set to `0.2`. Some features can be excluded from drift tracking using the **excluded features** options.

- The **Feature Importance** threshold allows users to define the threshold that differentiates the most important features from others is. In *Figure 13.13*, `0.45` is entered as the **Permutation Importance** metric threshold to achieve this goal. By so doing, features with permutation importance over `0.45` are deemed as **Failing**, while those with lower importance are considered **At Risk**. Here, some features are seen as important irrespective of the actual feature importance, and these can be selected using the **starred features** options.

- The **At Risk** and **Failing** thresholds alternatively enable the configuration of the minimum number of low- and high-importance features that are necessary for sending **At Risk** or **Failing** notifications. The rule present in *Figure 13.13* allows the following:

  a. At-risk notifications to be sent when two or more low-importance features (with **Permutation Importance** of less than `0.45`) drifts beyond a **PSI** of `0.2`

  b. Failing notifications to be sent when five or more low-importance features have significant drift

  c. Failing notifications to be sent when one or more high-importance features have a drift whose **PSI** exceeds `0.45`

4. Notifications pertaining to the **Accuracy** production model need to be set up within the monitoring window (as shown in *Figure 13.13*). Here, the **Metric** accuracy, **Measurement** threshold, and rules for **At Risk** and **Failing** notifications are defined:

   a. Because the deployment is based on a classification problem, its **Metric** accuracy is selected from classification options. These include `AUC`, `Accuracy`, `Balance Accuracy`, `LogLoss`, and `FVE Binomial`, among others. In this case, `LogLoss` is chosen.

   b. The **Measurement** option defines how changes in the accuracy metric between production prediction and training data are compared. Here, the `percent` change is selected.

   c. Rules are then set for **At Risk** and **Failing** notifications to be sent. In this case, **At Risk** notifications are sent when the accuracy of the model for prediction data is 5% that of the training. Similarly, 10% is the threshold that triggers a **Failing** notification.

   d. As with **Data Drift**, the **Accuracy** notifications are set to be sent `Every day` at `1:00`. These could be configured to any cadence between daily and quarterly.

5. After this setup, clicking the **Save new setting** button activates the notification routine. However, it is noteworthy that any stakeholder who has access to the deployment can configure the notifications they want to receive. When models' changes become significant, it might become necessary to replace the model in the deployment.

    The performance of production models tends to decay with time. This raises the need for models to replace those in a deployment. Within the MLOps offerings, DataRobot offers the model replacement functionality. To change the model in a deployment, you navigate to the **Deployment Overview** window. The **Replace model** option is selected from the **Action** button on the right-hand side of the **Deployment Overview** window (see *Figure 13.14*):

6. Clicking the **Replace model** option presents a **Paste DataRobot model URL** request. This URL is for the location at which the new model can be found when opened from the leaderboard:

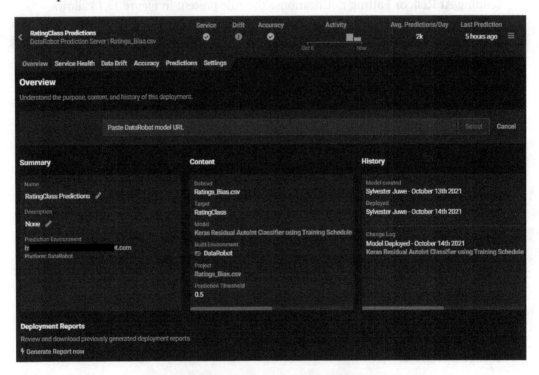

Figure 13.14 – Production model replacement

7. When **Select** is clicked, there is a prompt for the rationale for the model replacement. For this, options for **Replacement Reason** include Accuracy, Data Drift, Errors, Scheduled Refresh, and Scoring Speed. As shown in *Figure 13.15*, Data Drift is selected in this case.

8. Finally, **Accept and replace** is clicked:

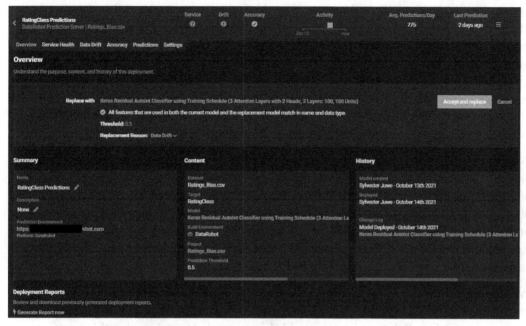

Figure 13.15 – Selecting the rationale for model replacement

Having replaced the model in the deployment, future predictions from this deployment will use the updated model. It is important to highlight that model replacement can only be carried out by the deployment owner. There are situations when the commercial impact of models is significant. In such situations, it is advisable to test the new or challenger model in a synthetic or simulated environment before switching models. In the typical data science workflow, the champion/challenger model scenario is well established. Here, challenger models compute predictions and their performance is compared with the one in production, the champion. With the testing and impact analyses complete, we are now ready to deploy our model. DataRobot provides data scientists the ability to test multiple challenger models while the champion is in production. This simplifies the model selection process when a model is to be replaced.

MLOps also offers the capability for changes in the model to be reviewed by different stakeholders. For this to happen, models are assigned importance levels as part of their deployments. These importance levels depend on the strategic commercial impact the model outcomes have on the business, the volume of predictions, and regulatory expectations. The importance levels thereafter drive who needs to review changes of the deployments before they are implemented.

Summary

In this chapter, we highlighted the value of establishing a framework guiding the use of ML models in businesses. ML governance capability supports users in ensuring that ML models continue to deliver commercial value while meeting regulatory expectations. Also, we set controls for what different levels of stakeholders can do with ML deployments. In some industries, there is a need to seriously consider the impact of bias in any decision process. Because ML models are based on data that might have been affected by human bias, it is possible that these models will compound such bias. As such, we explored ways to mitigate ML bias during and after model development.

We also examined the effects features have on the outcome variable. Such changes could have a critical bearing on business outcomes, hence the need to monitor the performance of model outcomes in production. During this chapter, we explored ways the performance of models could be assessed over time. Importantly, we learned how to configure notifications when there are significant changes in data drift or/and model accuracy. Additionally, we examined how a model in production could be switched to a challenger as needed.

We also highlighted some other MLOps features that were not covered in depth as part of this chapter. In the next chapter, we are going to look at what we think the future holds for DataRobot and automated ML in general. Also, given that this book is not all-encompassing with regards to DataRobot and the platform keeps expanding its capabilities, in the next chapter, we will point out some places where additional information for further development could be accessed.

14
Conclusion

In the preceding chapters, we learned to build and deploy models with DataRobot. We learned the basics, along with some advanced data science concepts. The disciplines of ML and data science continue to evolve rapidly. In response, tools such as DataRobot are also enhancing their capabilities. While these tools will continue to advance and make data scientists more productive, it is also expected that the role of good data scientists will become increasingly important and better understood. I hope we have convinced you that data scientists are not about to become obsolete and also that these tools have a lot to offer for data scientists, regardless of their level of expertise. It is also true that the methods and tools that we have right now are still very limited when it comes to intelligence. There is a lot of learning and uncovering to be done before the systems we build can be called intelligent.

The information we have provided so far should help you get started in developing and deploying models and start impacting your business. There is still a lot more to learn in this journey. In this chapter, we will provide sources for finding additional information regarding DataRobot and also discuss what the future of data science and DataRobot might look like. We're going to cover the following main topics:

- Finding out additional information about DataRobot
- Future of automated machine learning
- Future of DataRobot

Finding out additional information about DataRobot

There are many sources for additional information on ML, AI, and related methods. The best sources for DataRobot-related information are the following:

- **DataRobot Website**: `https://www.datarobot.com`

    The website contains information about the platform and several resources and case studies. It also contains links to other useful sites, some of which are described next.

- **DataRobot Community Site**: `https://community.datarobot.com`

    This is a site for the DataRobot user community. You can go to this site to connect with other users, see what information they are discussing, and ask questions.

- **DataRobot Community GitHub**: `https://github.com/datarobot-community`

    This is a GitHub site with repositories of DataRobot-related projects. You can see many code samples and examples for a wide range of DataRobot-related tasks. Here, you will find Python scripts and notebooks as well as samples in many other languages. This is a great place to go if you are ready to start using the DataRobot API.

- **DataRobot Python Client**: `https://datarobot-public-api-client.readthedocs-hosted.com/en/v2.25.0/`

    This link provides information about the publicly available Python package for DataRobot. Here, you will find information on how to get this package, install it, and use it. Note that you need a valid license to make use of this package.

- **DataRobot Documentation**

    Once you are logged in to DataRobot, you can see links to extensive DataRobot documentation for the tool itself as well as the DataRobot API, as we discussed in *Chapter 1*, *What Is DataRobot and Why You Need It*.

- **DataRobot Documentation Website**: `https://docs.datarobot.com`

    DataRobot recently decided to release all documentation on this site. This includes the DataRobot platform, APIs, tutorials, and notebooks, as well as a glossary. This is now the default site to look for all DataRobot-related documentation.

There are many publicly available information sources for ML methods, AutoML, and AutoML tools that you can find and explore online. Let's now discuss where ML seems to be headed.

Future of automated machine learning

It has been over 5 years since automated ML tools started appearing in the data science community. There are now several open source tools (for example, TPOT and Auto-WEKA) as well as proprietary tools (for example, Kortical and H2O) on the market. All major cloud providers now have some AutoML offering. Interest in these tools has been rising. It is safe to say that interest in these tools will keep rising and more offerings should be expected to come onto the market. We expect the tools to keep expanding the scope of tasks that will be automated to cover more aspects of the model development process. We can already see tools such as DataRobot offering more functions through internal development as well as through acquisitions such as Paxata. While it remains true that these tools do not support all use cases or types of modeling, they do cover a large number of use cases in a typical organization. More capabilities and algorithms are being added every quarter. We also expect that some niche vendors will emerge that focus on specific problems or methods such as reinforcement learning.

Many of the ideas and methods described in this book can be used with other AutoML tools or even with just a notebook environment with open source Python libraries. The AutoML tools essentially serve to automate many of the mundane and labor-intensive tasks. Even with these developments, we expect that not all aspects of the model development process can be automated. We covered many of these aspects in different chapters of this book. Hence, it is our belief that trained and experienced data scientists will always be needed. Adoption of these tools will enable the data scientists to cover more use cases and the resulting models will be of higher quality. These tools will also expand the reach of advanced algorithms to analysts in organizations who understand data science concepts but are not as familiar or comfortable with programming. Like any tool or technology, there is the possibility of misuse. Data-savvy organizations will put model governance and training programs in place to prevent problems relating to the use of bad or biased data, solving the wrong problem, or creating solutions that are not actionable. Also, the organizations should look into their specific needs and then select the tools most appropriate for their situation.

It is also expected that ML will need to move beyond pattern matching and we expect a bigger focus on causal modeling in the coming years. This is because many industrial use cases require specific decisions to be made or interventions that require decision makers to understand causal impacts and the consequences of these decisions. We have discussed some of these topics and methods. These methods defy automation in their current form and require substantial human input. For these methods to become mainstream, aspects of these methods need to be automated to enable organizations to start adopting them.

As the tools expand to include causal modeling, it will become imperative that capabilities be added to build models at a systemic level or be able to compose models to create a broader view of the systems under consideration. We will see the emergence of "digital twin" models that will represent entire systems. It will also be necessary to manage and control the configuration of many experiments that will need to be conducted. Given that large datasets are involved in building single models, current methods for running multiple experiments on multiple models are not scalable.

We also expect more data scientists to adopt the AutoML tools as they start realizing that these tools will make them more productive, while still giving them all the flexibility they need. This is done by combining programming languages, APIs, and automation support.

Another trend to watch is tool vendors' focus on solving customers' problems. It is very common for some vendors to lose customer focus as they grow bigger, which ultimately can lead to their downfall. This is not a new phenomenon; it has been happening to tool vendors for a long time. We have seen many vendors lose this focus and fade into oblivion. Some of this will also happen to AutoML vendors, so be on the lookout for vendors who are not very responsive to your needs.

Future of DataRobot

DataRobot was the early pioneer in the AutoML space and seems to be the dominant player, but there are many others (H20, Kortical, and Google Cloud AutoML, to name a few) that are catching up rapidly. Many of the large cloud players are jumping into this space and have offerings that are very attractively priced. DataRobot continues to offer additional capabilities combined with good support from experienced data scientists. To that end, we expect that the DataRobot API will continue to evolve and become more robust to allow experienced data scientists to use DataRobot in a highly flexible and automated way.

We have noticed new capabilities being released even as this book is being written, such as the recent acquisition of the Zepl notebook platform. In addition to that, DataRobot continues to acquire other companies to round out its offerings. Recently, a lot of focus has been on MLOps and enabling the rapid deployment of models. As the features and capabilities increase, the learning curve for understanding and using DataRobot also increases. It is our experience that many data scientists who use DataRobot are not fully aware of its capabilities and are not utilizing it to its fullest extent. We hope that books such as this will help with this issue. This alone, however, will not be enough to maintain a competitive position. Expanding into next-generation capabilities outlined in the previous section will be necessary to maintain an edge.

In conclusion, we hope that we have given you a good overview of how you can put DataRobot to practical use in your organization right away and make your data science journey more productive. Hopefully, we have convinced you that a data scientist's role is not just to produce predictions, but to help enable good decisions. Just as a lot of what we do automates the tasks of many roles in the organization, it is also imperative that we do not fear the automation of data science tasks. We hope that this book provides useful ideas to a broad range of data scientists across a broad range of industries.

`Packt.com`

Subscribe to our online digital library for full access to over 7,000 books and videos, as well as industry leading tools to help you plan your personal development and advance your career. For more information, please visit our website.

Why subscribe?

- Spend less time learning and more time coding with practical eBooks and Videos from over 4,000 industry professionals

- Improve your learning with Skill Plans built especially for you

- Get a free eBook or video every month

- Fully searchable for easy access to vital information

- Copy and paste, print, and bookmark content

Did you know that Packt offers eBook versions of every book published, with PDF and ePub files available? You can upgrade to the eBook version at `packt.com` and as a print book customer, you are entitled to a discount on the eBook copy. Get in touch with us at `customercare@packtpub.com` for more details.

At `www.packt.com`, you can also read a collection of free technical articles, sign up for a range of free newsletters, and receive exclusive discounts and offers on Packt books and eBooks.

Other Books You May Enjoy

If you enjoyed this book, you may be interested in these other books by Packt:

Engineering MLOps

Emmanuel Raj

ISBN: 978-1-80056-288-2

- Formulate data governance strategies and pipelines for ML training and deployment
- Get to grips with implementing ML pipelines, CI/CD pipelines, and ML monitoring pipelines
- Design a robust and scalable microservice and API for test and production environments
- Curate your custom CD processes for related use cases and organizations
- Monitor ML models, including monitoring data drift, model drift, and application performance
- Build and maintain automated ML systems

Machine Learning with Amazon SageMaker Cookbook

Joshua Arvin Lat

ISBN: 978-1-80056-703-0

- Train and deploy NLP, time series forecasting, and computer vision models to solve different business problems

- Push the limits of customization in SageMaker using custom container images

- Use AutoML capabilities with SageMaker Autopilot to create high-quality models

- Work with effective data analysis and preparation techniques

- Explore solutions for debugging and managing ML experiments and deployments

- Deal with bias detection and ML explainability requirements using SageMaker Clarify

- Automate intermediate and complex deployments and workflows using a variety of solutions

Packt is searching for authors like you

If you're interested in becoming an author for Packt, please visit `authors.packtpub.com` and apply today. We have worked with thousands of developers and tech professionals, just like you, to help them share their insight with the global tech community. You can make a general application, apply for a specific hot topic that we are recruiting an author for, or submit your own idea.

Share Your Thoughts

Now you've finished *Agile Machine Learning with DataRobot*, we'd love to hear your thoughts! Scan the QR code below to go straight to the Amazon review page for this book and share your feedback or leave a review on the site that you purchased it from.

https://packt.link/r/1801076804

Your review is important to us and the tech community and will help us make sure we're delivering excellent quality content.

Index